T0302336

Fundamentals of Nanoindentation and Nanotribology IV

MATERIALS RESEARCH SOCIETY
SYMPOSIUM PROCEEDINGS VOLUME 1049

Fundamentals of Nanoindentation and Nanotribology IV

Symposium held November 26–29, 2007, Boston, Massachusetts, U.S.A.

EDITORS:

Eric Le Bourhis
Université de Poitiers
Futuroscope-chasseneuil, France

Dylan J. Morris
National Institute of Standards and Technology
Gaithersburg, Maryland, U.S.A.

Michelle L. Oyen
Cambridge University
Cambridge, United Kingdom

Ruth Schwaiger
Forschungszentrum Karlsruhe
Karlsruhe, Germany

Thorsten Staedler
Universitaet Siegen
Siegen, Germany

Materials Research Society
Warrendale, Pennsylvania

CAMBRIDGE
UNIVERSITY PRESS

University Printing House, Cambridge CB2 8BS, United Kingdom

One Liberty Plaza, 20th Floor, New York, NY 10006, USA

477 Williamstown Road, Port Melbourne, VIC 3207, Australia

314-321, 3rd Floor, Plot 3, Splendor Forum, Jasola District Centre, New Delhi - 110025, India

79 Anson Road, #06-04/06, Singapore 079906

Cambridge University Press is part of the University of Cambridge.

It furthers the University's mission by disseminating knowledge in the pursuit of education, learning and research at the highest international levels of excellence.

www.cambridge.org
Information on this title: www.cambridge.org/9781558999893

Materials Research Society
506 Keystone Drive, Warrendale, PA 15086
http://www.mrs.org

© Materials Research Society 2008

This publication is in copyright. Subject to statutory exception and to the provisions of relevant collective licensing agreements, no reproduction of any part may take place without the written permission of Cambridge University Press.

This publication has been registered with Copyright Clearance Center, Inc. For further information please contact the Copyright Clearance Center, Salem, Massachusetts.

First published 2008
First paperback edition 2012

Single article reprints from this publication are available through University Microfilms Inc., 300 North Zeeb Road, Ann Arbor, MI 48106

CODEN: MRSPDH

A catalogue record for this publication is available from the British Library

ISBN 978-1-558-99989-3 Hardback
ISBN 978-1-107-40859-3 Paperback

Cambridge University Press has no responsibility for the persistence or accuracy of URLs for external or third-party internet websites referred to in this publication, and does not guarantee that any content on such websites is, or will remain, accurate or appropriate.

This work was performed under the sponsorship of the U.S. Dept. of Commerce, National Institute of Standards and Technology, Cooperative Agreement No. 60NANB7D6172.

CONTENTS

NANOMECHANICS, TRIBOLOGY AND NANOSTRUCTURES

SIZE EFFECTS AND INDENTATION OF THIN FILMS

*Invited Paper

*Invited Paper

vi

PREFACE

Symposium AA, "Fundamentals of Nanoindentation and Nanotribology IV," was held November 26–29 at the 2007 MRS Fall Meeting in Boston, Massachusetts. This symposium was the fourth of a series highlighting emerging topics in nanoindentation and nanotribology, including the development of new methods for characterizing nanoscale mechanical and tribological properties.

Nanoindentation and nanotribology are fundamental, evolving, and complementary disciplines within materials science. In recent years there has been rapid convergence of the biomechanical and materials disciplines, as well as explosive growth of the nanotube and nanostructured materials fields. The expansion of nanomechanical testing into these new fields has been accompanied by similarly rapid growth in our understanding of, and ability to perform, mechanical tests with ever-smaller forces and displacements. This occurs even as the materials and relevant length scales diverge from traditional engineering materials. Furthermore, the understanding of fundamental mechanical measurement techniques must continue to advance to understand and design new systems and materials to meet the challenges of technology.

Symposium AA had nearly 100 presentations from participants from 13 countries. This volume contains many of the presentations from Symposium AA, including those from a joint session with Symposium OO, "Solids at the Biological Interface." Focused session topics in Symposium AA included *in situ* methods, nanotribology and nanostructures, modeling and analysis of indentation, the nanomechanics of polymers, tribology of biological materials, size effects and thin films, and temperature- or time-dependent indentation phenomena. We hope that this volume — a snapshot of the state-of-the-art in nanoindentation and nanotribology — will serve as a useful reference, and creative inspiration, for students, scientists, and engineers in the nanomechanical disciplines.

Eric Le Bourhis
Dylan J. Morris
Michelle L. Oyen
Ruth Schwaiger
Thorsten Staedler

February 2008

ACKNOWLEDGMENTS

The symposium organizers acknowledge with gratitude the financial support provided by CNRS Saint Gobain, FEI Electron Optics BV, Hysitron Inc., MTS Nano Instruments and the National Institute of Standards and Technology.

We greatly appreciate the excellent administrative and technical support provided by the MRS staff, which was a major contribution to the success of this symposium.

We would like to thank the session chairs for doing an excellent job in keeping the program on schedule, and for moderating stimulating discussion throughout.

Finally, we are most appreciative to the community of authors for submitting their latest research results and advances for publication in this volume. We owe much gratitude to peer-reviewers whose diligent and thorough reviews made these proceedings possible.

MATERIALS RESEARCH SOCIETY SYMPOSIUM PROCEEDINGS

MATERIALS RESEARCH SOCIETY SYMPOSIUM PROCEEDINGS

Prior Materials Research Society Symposium Proceedings available by contacting Materials Research Society

Nanomechanics, Tribology and Nanostructures

Mater. Res. Soc. Symp. Proc. Vol. 1049 © 2008 Materials Research Society   1049-AA02-04

Comprehensive Mechanical and Tribological Characterization of Ultra-Thin-Films

Norm Gitis, Michael Vinogradov, Ilja Hermann, and Suresh Kuiry
Center for Tribology, Inc., 1715 Dell Ave, Campbell, CA, 95008

ABSTRACT

Mechanical and tribological properties such as hardness, Young's modulus, friction, and scratch adhesion strength of various coatings and ultra-thin films are reported. These results, obtained using a Universal Nano+Micro Tester UNMT-1, indicate that a substrate effect for ultra-thin films is substantial when using conventional static nanoindentation technique, while negligible with an advanced dynamic nano-indentation. Comparative results of hardness and Young's modulus obtained from various techniques are presented. A means to evaluate friction and adhesion strength of thin films is highlighted, using DLC specimens as an example.

INTRODUCTION

Evaluation of mechanical and tribological properties of bulk and coating materials is of great importance for design and development of engineering components with enhanced structural and wear performance [1-3]. Traditional indentation tests of bulk materials include macro-hardness measurements at high loads of the order of kN. Micro-hardness measurements of coatings and bulk materials are usually performed under loads of the order of N. As the industrial technology advanced, the characterization technique for mechanical properties of thin films and coating shifted its range to mN and even μN.

In recent years, nano-indentation has become a popular technique for evaluation of hardness and Young's modulus of films and coatings. The nature of stress distribution in the front of a nano-indenter tip makes the indentation vulnerable to a substrate effect. Usually, the indentation depth should be restricted to 5 - 10% of the film thickness to limit the stress field within the thickness of the film and thus, to avoid the substrate influencing the results of such measurements. Such measurements at ultra-shallow depths would require an extremely high accuracy of both depth monitoring and tip calibration. Although AFM-based nano-indentation shows some potential to circumvent such problem, the use of compliant tips limits its applications to soft materials. A novel Nano-analyzer enables quantitative characterization of ultra-thin films, including hard and super-hard films, at shallow depths with a negligible substrate effect.

Primary objective of the present investigation is to compare different techniques to perform mechanical and tribological tests on micro and nano-level using a single tester.

EXPERIMENTAL

The micro-indentation, micro-scratch, nano-indentation, nano-scratch, and nano-imaging techniques were employed for evaluation of mechanical and tribological properties of numerous bulk, coating and film specimens using the same Universal Nano+Micro Tester model UNMT-1, designed and manufactured by CETR, Inc. The photograph of its nano-head is presented in Figure 1.

UNMT-1 has several easily interchangeable modules for precision tests of hardness, Young's modulus, friction, and adhesion of bulk, coating and film materials, including:
- traditional micro-indentation up to a load of 1.2 kN, with Rockwell, Vickers, and Knoop indenters according to ASTM E18-05, ASTM E92-82, and ASTM E384-99 standards,

respectively. It can perform post-test measurements of diagonals of indentation with an optical microscope as required by Vickers and Knoop hardness test procedures, and of indentation depth with a capacitance sensor for Rockwell hardness test,

- instrumented micro-hardness tests per the ISO 14577-1/02 up to a load of 1.2 kN with the same Rockwell, Vickers, or Knoop indenters, but *in-situ* monitoring of load and depth and automatic calculations of both instrumented hardness and Young's modulus,

- instrumented static nano-indentation tests per the ISO 14577-1/02 (loads from 0.1 μN to 0.5 N) with Berkovich, conical and cube-corner indenters, *in-situ* monitoring of load and displacement and automatic calculations of both instrumented hardness and Young's modulus,

- instrumented dynamic Young's modulus tests with spherical, Berkovich, and other indenters, *in-situ* monitoring of tip frequency changes during surface scanning, and both calculations and maps of Young's modulus,

- micro-scratch-hardness tests per the ASTM G171-03 (loads from centi-N to hecto-N, sliding distances from a few microns to many millimeters) with numerous indenters (spherical, conical, micro-blades, etc.), under a constant load or any other loading profile with capability of simultaneous monitoring of friction, acoustic emission, electrical resistance, etc.

- nano-scratch-hardness tests per the ASTM G171-03 (loads from 0.1 micro-N to hecto-N, sliding distances from 1 to 100 microns) with numerous indenters (Berkovich, spherical, etc.) and AFM-like imaging of scratches with the same tip.

The UNMT-1 allows for multi-scale measurements of the same sample without its removal, just with an easy exchange of the mentioned indentation and scratch modules.

RESULTS AND DISCUSSION

The evaluation results of mechanical and tribological properties of numerous specimens in ambient conditions are summarized in Table 1. Whereas the "+"shows that the technique was sufficient to measure the property of the sample, the "−"indicates that the technique failed to evaluate the film properties without the substrate effect.

Table 1: Summary of hardness tests performed using UNMT-1					
Specimens	Micro-indentation		Instrumented		
	Traditional	Instrumented	Micro-scratch	Nano-indentation	Nano-scratch
Bulk Materials	+	+	+	+	+
20μm metallic film	−	+	+	+	+
2μm metallic film	−	−	+	+	+
4 nm DLC film	−	−	−	−	+

Micro-indentation

Instrumented indentation tests were performed on bulk metal specimens with a Rockwell diamond indenter with tip radius of 200 micron. The maximum loads for aluminum, brass, and steel specimens were about 30, 70 and 150 N, respectively. Figure 2 shows the representative load-displacement plots, the unloading portions of which were analyzed using the Oliver-Pharr methodology [4]. Table 2 presents the mean hardness and Young's modulus values, obtained from 20 tests on each specimen. It also shows the average value of 20 hardness data obtained from traditional Vickers hardness test on the same samples. The average results of the instrumented and traditional techniques were practically the same (the 10-time difference is due to the Vickers scale), while the deviation in hardness values were found to be less in instrumented indentation compared to the traditional Vickers hardness test.

Fig. 1. Nano-indentation module for UNMT-1   **Fig. 2.** Load-displacement plots for metals

Micro-Scratch

The instrumented micro-scratch tests included sliding-scratching over 5 mm length at a speed of 0.2 mm/s under a constant load, using a Rockwell diamond indenter, on test specimens and a reference material. The scratching loads for the specimen (F_S) and the reference (F_R) should be such that the scratch widths should be similar. The scratch hardness (HS) is:

$$HS = H_R \cdot \frac{F_S}{F_R} \left(\frac{W_R}{W_S} \right)^2 \qquad \qquad \text{.........(1)}$$

H_R is the hardness of the reference material, W_R and W_S are the scratch widths on the reference and specimen, respectively. Table 3 shows the scratch width and hardness on the fused silica, bare and coated metal; the coating exhibited hardness three times higher than the substrate.

Table 2: Hardness and modulus for aluminum, brass and steel specimens			
Specimen	Hardness, MPa	Young's modulus, GPa	Vickers hardness
Aluminum	968±23	74±2	97±11
Brass	2004±19	127±2	201±9
Steel	5202±47	195±8	519±20

Table 3: Scratch hardness on fused silica, bare and coated specimens								
Sample	Load, N	Scratch width, µm						Hardness,
		1	2	3	4	5	Mean	GPa
Fused Silica	0.2	9.96	9.78	9.68	10.23	10.03	9.94	9.5
Coating	0.1	7.74	7.79	7.83	8.23	7.98	7.91	7.5
Substrate	0.1	13.75	13.64	13.65	13.86	14.05	13.79	2.4

Static and Dynamic Nano-indentation

Instrumented nano-indentation was performed in both static and dynamic modes, using the same Berkovich indenter. Figures 3 and 4 show ten load-displacement curves obtained from static nano-indentation tests on a fused silica and 2-µm polymer coating on silicon. Figure 5 presents ten frequency-approach curves obtained from dynamic nano-indentation on reference polycarbonate specimen and on the 2-µm polymer film on silicon. One can see excellent data repeatability of both sets of data.

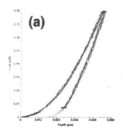 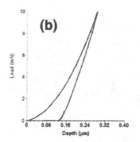

Fig. 3. Load-displacement curves on fused silica up to (a) 0.4 mN and (b) 10 mN.

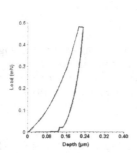 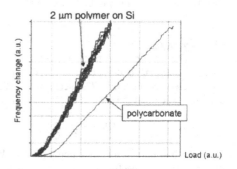

Fig. 4. Load-displacement curves into 2 µm polymer on Si.

Fig. 5. Frequency-approach curves from dynamic nano-indentation.

Table 4 summarizes Young's modulus data from static and dynamic nano-indentation tests on various specimens. The static nano-indentation showed good results for bulk materials and micron-thick coatings, but a significant substrate effect for nano-coatings. The dynamic nano-indentation showed good results for all the specimens, including for ultra-thin films.

Nano-Scratch

The UNMT-1 Nano-analyzer module was used to measure hardness by nano-scratching, followed by AFM-like nano-imaging with the same tip. Its software allows for image analysis to obtain scratch depth profile along any direction of the nano-image. The hardness values were obtained from equation (1) above.

Table 4: Young's modulus data from static and dynamic nano-indentation		
Specimens	Young's Modulus, GPa	
	Nano-indenter	Nano-analyzer
Silicon 100	164±18	164±14
12 µm polymer on Si	6.85±0.07	6.0±0.5
2 µm polymer on Si	7.36±0.31	6.0±0.5
2 µm Ti on Si	98±13	98±10
2 µm DLC on Si	381±24	382±19
1 µm DLC on Si	372±29	379±21
100 nm DLC on Si	314±32*	370±25
4 nm DLC on Si	195±39*	361±27
Reference Polycarbonate	3.62±0.06	3.50±0.04
Reference Fused Silica	71.20±0.65	72.9±0.8
*Substrate effect		

Figure 6 shows an image of a 15 x 15 µm area of a 12 µm polymer coating, scratched under loads of 0.5, 0.2 and 0.1 mN (left to right). The scratch depth profiles along the X-X direction of Figure 6 are shown in Figure 7.

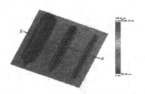

Fig.6. Nano-image of 3 nano-scratches   **Fig.7.** Three repeated scratch depth profiles

Table 5 summarizes data of nano-indentation *vs.* nano-scratch tests. The nano-indentation showed good results for micro-coatings, but a substrate effect (denoted as *) for nano-coatings. Such substrate effect was absent in the nano-scratch tests, when a stress distribution in the front of a moving indenter stays within the coating and does not extends into the substrate.

Table 5: Nanoindentation and Nano-scratch hardness data		
Specimens	Hardness, GPa	
	nano-indentation	nano-scratch
Silicon 100	11.4±1.8	11.6±1.0
12 µm polymer on Si	0.29±0.01	0.27±0.03
2 µm polymer on Si	0.44±0.01*	0.33+0.04
1 µm DLC on Si	30.5±2.9	31.2±1.7
100 nm DLC on Si	24.6±3.6*	30.8±1.7
4 nm DLC on Si	15.7±3.2*	29.1±1.9
Reference Polycarbonate	0.23±0.004	0.21±0.03
Reference Fused Silica	9.52±0.07	9.57±0.50

Scratch-Adhesion

Scratch-adhesion tests with a micro-indenter or a micro-blade are performed by sliding under a linearly increasing load (Fz). A failure of the coating is characterized by sudden change in coefficient of friction (COF) or contact high-frequency acoustic emission (AE). Figure 8 shows data from three scratch-adhesion tests with a diamond tip of 12.5 μm radius on a patterned LCD display with layers of indium-tin-oxide/overcoat/matrix on a glass substrate. At the load of about 16 mN, both COF and AE increased, indicating the failure. The periodic bumps in the COF and Fz plot were due to the rapid interactions of the tip with the patterned specimen. Table 6 shows the COF, AE and scratch-adhesion strength data of the top indium-tin-oxide layer.

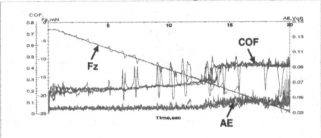

Fig.8. Fz, COF, and AE plots for scratch adhesion tests on LCD specimen.

Table 6: Scratch-adhesion data on LCD specimen				
Adhesion Strength, mN	COF		AE	
	Before failure	After failure	Before failure	After failure
16.2	0.242	0.410	0.04	0.08
16.5	0.242	0.411	0.04	0.06
16.2	0.237	0.417	0.04	0.09
Mean : 16.3	0.240	0.413	0.04	0.08

CONCLUSIONS

The repeatable substrate-independent results were obtained in the indentation tests with the indents under 5-10% of film thickness and in the scratch tests with the scratches under 30-35% of film thickness. In the micro-indentation tests on metals, traditional Rockwell and Vickers hardness tests produced more data variability than the instrumented-hardness tests, though the statistics requires more data. The UNMT-1 provides a unique single platform for comparative studies of mechanical and tribological properties on micro- and nano-levels.

FUTURE STUDY

Next tests will be focused on indentation and scratch evaluation of coatings at extreme high and low temperatures, using UNMT-1 chamber modules.

REFERENCES

1. D. Tabor, Phil. Mag., A74 (5), 1207 (1996).
2. N. Gitis, J. Xiao, and M. Vinogradov, J. ASTM Int. STP 1463, 2 (9), 80(2005).
3. N. Gitis, etc., Proc. Int. Joint Trib. Conf., San Diego, IJTC2007-44025 (2007)
4. W. Oliver and G. Pharr, J. Mat. Res., 19(1), 1(2004); 7, 1564(1992).

Mater. Res. Soc. Symp. Proc. Vol. 1049 © 2008 Materials Research Society 1049-AA02-07

Mechanical response of a single and released InP membrane

Eric Le Bourhis[1], and Gilles Patriarche[2]
[1]Laboratoire de Métallurgie Physique, Université de Poitiers, UMR 6630 CNRS, SP2MI-Téléport 2-Bd Marie et Pierre Curie, BP 30179, Futuroscope-chasseneuil Cedex, F-86962, France
[2]Laboratoire de Photonique et de Nanostructures, CNRS UPR 20, Marcoussis, 91460, France

ABSTRACT

InP membranes have been microfabricated and released on top of an InP substrate. The release process comprises membrane photolithography, interlayer sacrificial etching of a InP/InGaAs/InP substructure and drying with CO_2 at critical point. The mechanical response of the obtained small (40 μm) and thin (0.4 μm) membranes could be tested by nanoindentation. While keeping their epitaxial orientation through the fabrication process, delamination of the membrane was observed to occur before the indenting load reached 10 mN. Then cracking of the membrane was detected as a pop-in on the loading curves for loads larger than 10 mN.

INTRODUCTION

Mechanical behavior of single objects has attracted much interest in the past few years [1]. In this field, semiconductor defect and size engineering allows fabricating model structures. Moreover, indentation technique has proved to be powerful to test small volumes [2]. Originally, contact mechanics was developed for semi-infinite half space [3], this assumption being not fulfilled when the size of the plastic zone becomes of the order of one of the dimensions of the object [4-9]. 'Small' structures are expected to show a mechanical behavior deviating from that of a bulk. So far, length scale induced changes in the response of single semiconductor objects has been poorly addressed. Only recently were reported nanoindentation studies of GaAs single lines and of focused-ion beam (FIB) milled GaAs pillars in the μm-length scale [7-9]. Therefore, we decided to investigate the mechanical response of individual, small (40 μm in diameter) and thin (~ 0.4 μm) InP membranes. These were fabricated from InP/InGaAs/InP substructures and released on top of the InP substrate. Nanoindentation could be performed on these single membranes under increasing loads while interferential and transmission electron microscopies allowed getting deeper insight into the deformation of such small objects.

EXPERIMENT

Undoped (001) InP substrates were used for the study. InP membranes were micro fabricated and released on top of the InP substrate in a four step process. The first step was to grow an InP/InGaAs heterostructure on an InP substrate by low-pressure metal organic vapor phase epitaxy (MOVPE). The InGaAs layer was meant to be sacrificial and etched subsequently in order to release the membrane. Before this third step, the membranes were fabricated by photolithography of a poly-methyl methacrylate (PMMA)-based resin film previously deposited on top of the InP/InGaAs coated InP substrates. After the revealing step, a-SiN was deposited by plasma-enhanced chemical vapor deposition (PECVD). A lift-off process allowed producing a-SiN masks on the InP surface. Dry etching was then carried out with $SiCl_4$ plasma using a reactive-ion-etching (RIE) machine. The a-SiN masks were then removed from the InP/InGaAs/InP substructures using a HF solution (end of the second step). The InGaAs sublayer was etched preferentially and the obtained structure later dried by CO_2 at critical point and released on the substrate (third and fourth steps). The thickness of the membranes was measured by profilometry using a DEKTAK 3 ST from VEECO. The membranes under study were determined to be about 400 nm thick while their diameter was about 40 µm. They were deformed by a Berkovich diamond pyramid using a NHT machine from CSEM (Switzerland) employing the X-Y tables to position the relevant area (membranes) under the tip. The tests were performed either on the membranes or on the bare substrates (Figure 1b) in the force-control mode of the machine. The calibration procedure suggested by Oliver and Pharr [10] was used to correct for the load-frame compliance of the apparatus and the imperfect shape of the indenter tip. A phase-shift interferometer (Micromap 570 ATOS, Pfungstadt, Germany) was used to obtain quantitative images of the surface topography. To prepare TEM plan-view thin foils of the indented samples, the undeformed side of the samples (back side) was mechanically and chemically thinned with a bromine-methanol solution until sufficiently thin to transmit the electron beam. The samples (indented surface) were observed in a Philips CM20-Super-Twin microscope operated at 200 kV equipped with a double-tilt sample holder (±28°).

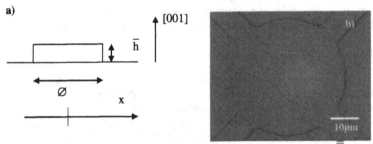

Figure 1: (a) Membrane geometric characteristics (diameter Ø, height $\bar{h}$). InP thin membrane surface is (001) while substrate is also (001) InP. (b) Membrane as indented under 10 mN. Note the crack appearing when the indent is made close to the membrane edge and only central indents were considered for the analysis.

RESULTS AND DISCUSSION

Figure 2 shows the loading and unloading curves obtained in the central part of a membrane under a maximum load F_{max} of 10 mN and at the reference surface (bare substrate). Slight differences on the indentation response are observed both on loading and unloading. The loading and most of the unloading curves are shifted to the right-hand side of the figure (larger penetrations) when the test is carried out on the central part of the membrane. The lowest portions of the unloading curves obtained on the reference surface and the membrane cross each other with a larger apparent recovery obtained in the case of a membrane. The maximum penetrations h_{max} and residual penetrations h_r of the indenter under a same maximum load of 10 mN are determined to be 315 ± 3 nm and 190 ± 10 nm on the membrane (± standard deviation of 10 events) and 305 ± 3 nm and 195 ± 3 nm on the bulk respectively. It is important to note that all these values remain below membrane thickness (400 nm) indicating that the interface between the membrane and substrate is not reached by the indenter tip extremity. Nonetheless, we expect the substrate is plastically deformed as discussed in more detail below. Observing larger penetrations on loading when the test is carried out on the central part of a membrane suggests that the membrane-substrate system deforms slightly more than the bare substrate. Instead, the recovery upon unloading is more important for the membrane than for the bare substrate. Latter result is further analyzed in view of topographic images.

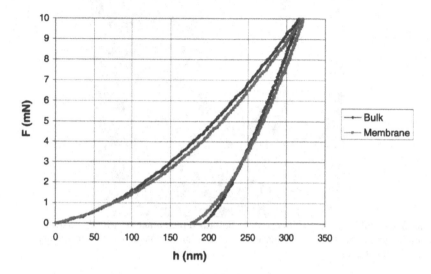

Figure 2: Load-penetration curves obtained on the central part of a released membrane and on the bulk under a maximum load of 10 mN.

Figure 3 shows the topography of the membrane surface after full unloading from 10 mN (ex-situ) and reveals humps at the indent sites with approximate height of 40 nm resulting from the membrane delamination from the substrate surface. It is to be noted that this 40 nm value is notably more than the difference observed on indentation curves and this discrepancy is to be attributed to the membrane rising after the tip is removed from the contact area. Instead, the indent sites at the bulk surface shows only slight marks under the same conditions and using the same technique. Buckling happens as indentation stress $\sigma_I = \dfrac{VE}{2\pi(1-v)\bar{h}a^2}$ exceeds Euler

buckling stress $\sigma_B = \dfrac{KE}{12(1-v^2)}\left(\dfrac{\bar{h}}{a}\right)^2$ [11-13]. Since we study an InP membrane released on an

InP substrate, residual stresses can be supposed to be null, V is the indentation volume, a the radius of the buckling crack, E and v the Young modulus and Poisson ratio respectively, K a coefficient ~14.7 [12]. It is interesting to note that, as the membrane thickness decreases indentation stress increases and buckling stress decreases making more favorable the delamination of the membrane. After indenting under a larger load of 20 mN, higher humps (~ 75 nm) are observed indicating further buckling of the membrane. Also, indentation loading curves reveal a new event characterized by a displacement excursion (so called pop-in) that is believed to be the signature of generation of radial cracks observed after full unloading by optical microscopy.

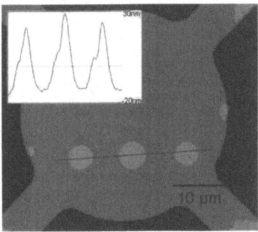

Figure 3. Optical interferential image of the deformed membrane showing humps after indentation under loads of 10 mN. Inset shows the height profile along the line passing through hump centres.

Figure 4 shows a TEM micrograph obtained after the deformation of the membrane under 10 mN maximum load. The indented zone reveals radial cracks that are expected to have appeared while preparing the TEM foil since they were not observed after full unloading by optical microscopy. As stated above, radial cracks appear above 10 mN maximum load. It is interesting to observe Moiré fringes in small zones at the vicinity of these cracks being signs of misorientation between the membrane and the substrate. Moiré fringes distance is about 60 to 80 nm, which represents a misorientation around the [001] direction of only 0.03-0.05°. On most observed areas however (and away from the cracks), we did not detect any Moiré fringes and determined the misorientation to be less than 0.01°. As stated before, the membrane is epitaxially grown with the continuation of the crystals from the substrate to the membrane. On removing the InGaAs interlayer and indenting under 10 mN, the orientation of the membrane with respect to the substrate is kept within a hundredth of degree.

The indented plastic zone is composed of (i) a central and dense region and (ii) rosette arms (RA) which extend along both <110> directions parallel to the indented surface as expected in a bulk specimen [14]. The central and dense zone corresponds to a highly strained zone with plastic strain reported as high as 28 % [2] and no individual dislocations being detected. The extension of this zone c can be predicted in a bulk specimen by Eq. (1) [2,14-16].

$$c = \sqrt{\frac{3}{2\pi} \frac{F_{max}}{Y}} \qquad (1)$$

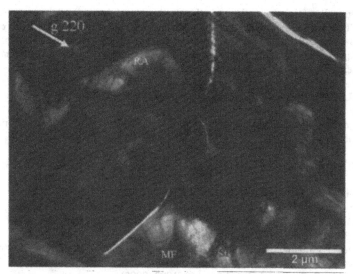

Figure 4: TEM plan view of the membrane indented under 10 mN, g = 220. MF, RA, SF are for Moiré fringes, rosette arm, stacking fault respectively.

In InP, the flow-stress value Y is about 1.3 GPa [14] and the central-plastic zone is expected to extend on c = 1.9 μm. This estimate is in good agreement with the observed extension in the membrane (Figure 4). Noticeably, both the membrane and the bulk substrate are of the same material ((001) oriented InP). The above result suggests that plastic deformation reaches the interface between the membrane and the substrate. The interface discontinuity between the released membrane and the substrate then acts as a barrier to plastic flow with the membrane bottom face being deformed [5,6]. The difference in plastic deformation between the membrane and the substrate drives an interfacial crack and yields membrane buckling as revealed by interferential microscopy. Noticeably, the substrate deforms plastically under the penetrations used here but material transfer from the membrane to the substrate is hindered. Importantly, the barrier is not resulting from a misorientation between the membrane and the substrate (which is less 0.01°) but from the employed process to release the membrane on top of the substrate. In fact, a much different behavior would be expected from a InP thin layer epitaxially grown directly on top of a InP substrate (with a crystal continuity from the substrate to the layer, e.g. homoepitaxy).

CONCLUSIONS

The mechanical response of thin InP membranes released on top of an InP substrate could be tested by nanoindentation. Elastic, plastic deformations, delamination and fracture could be detected gradually as load was increased. The membranes showed slightly larger deformations than a bare substrate and larger apparent recovery resulting from its delamination which occurs while plastic flow reaches the membrane-substrate interface. This interface acts as a barrier while no crystallographic misorientation between the membrane and the supporting substrate is detected by TEM resulting from the epitaxial growing of the substructure. Membrane cracking happens under higher loads and such events are correlated with displacement excursion (pop-in) on the loading curve. Work is in progress to get further insight into thickness dependence and release process influence on the membranes buckling.

ACKNOWLEDGMENTS

The authors would like to thank Dr. M. Strassner and Dr. I. Sagnes for the fabrication of the InP membranes.

REFERENCES

1. M.D. Uchic, D.M. Dimiduk, J.N. Florando, W.D. Nix, Science 305, 986 (2004).
2. E. Le Bourhis, *Glass Mechanics and Technology* (Wiley-VCH, Weinheim, 2007).
3. K.L. Johnson, *Contact Mechanics* (Cambridge University Press, Cambridge,1985).
4. Y. Choi, S. Suresh, Scripta Mater. 48, 249 (2003).
5. L. Largeau, G. Patriarche, E. Le Bourhis, A. Rivière, J.P. Rivière, Phil. Mag. 83, 1653 (2003).
6. G. Patriarche, L. Largeau, J.P. Rivière, E. Le Bourhis, J. Phys. D 38, 1140 (2005).
7. E. Le Bourhis, G. Patriarche, Acta Mater. 53, 1907 (2005).
8. E. Le Bourhis, G. Patriarche, Appl. Phys. Lett., 86, 163107 (2005).
9. J. Michler, K. Wasmer, S. Meier, F. Östlund, K. Leifer, Appl. Phys. Lett. 90, 043123 (2007).
10. W.C. Oliver, G.M. Pharr, J. Mater. Res. 7, 1564 (1992).
11. D.B. Marshall, A.G.J. Evans, Appl. Phys. Lett. 56, 2632 (1984).
12. P.R. Chalker, S.J. Bull, D.S. Rickerby, Mater. Sci. Eng. A 140, 583 (1991).
13. A.A. Volinsky, N.R. Moody, W.W. Gerberich, Acta Mater. 50, 441 (2002).
14. G. Patriarche, E. Le Bourhis, Phil. Mag. 82, 1953 (2002).
15. S. Harvey, H. Huang, S. Venkataraman, W.W. Gerberich, J. Mater. Res. 8, 1291 (1993).
16. D. Kramer, H. Huang, M. Kriese, J. Robach, J. Nelson, A. Wright, D. Bahr, W.W. Gerberich, Acta Mater. 47, 333 (1999).

Mater. Res. Soc. Symp. Proc. Vol. 1049 © 2008 Materials Research Society   1049-AA02-08

Indentation Response of Nanostructured Turfs

A. A. Zbib[1], S. Dj. Mesarovic[1], D. F. Bahr[1], E. T. Lilleodden[2], J. Jiao[3], and D. McClain[3]

[1]Mechanical and Materials Engineering, Washington State University, PO Box 642920, Pullman, WA, 99164-2920

[2]GKSS, Geesthacht, Germany

[3]Physics, Portland State University, Portland, OR, 97207-0751

ABSTRACT

When grown via chemical vapor deposition carbon nanotubes (CNTs) may take on the form of a "turf", consisting of many CNTs with a complex interconnectedness attached to an inflexible substrate. These turfs can be formed over large areas and with a range of heights (between 1 to 100 µm), and grown on photolithographically patterned catalysts to form different aspect ratios. This study focuses on the indentation and permanent deformation of CNT assemblages under applied contact loading. Nanoindentation was conducted on CNT turfs and the properties, nominally the turf's elastic modulus and hardness, were 14.9 MPa ± 5.7 MPa and 2 MPa respectively. The onset of permanent deformation during indentation occurred at applied stresses of 2.5 MPa. The turf's collective permanent deformation under applied compressive loading was also studied. A model predicting the buckling stress of CNT turfs is also described.

INTRODUCTION

A substantial amount of work in literature has been done investigating the mechanical[1], electrical[2] and thermal[3] properties of CNTs. However, most of this work has been focused on single tubes, whether single-walled carbon nanotubes or multi-walled carbon nanotubes. CNT turfs are complex structures of intertwined, nominally vertical tubes[4]. There are fewer reports that explore the properties of a bundle of CNTs in a turf like structure[5,6] and deformation under applied compressive loading[7,8]. Work conducted on CNT turfs has shown that these structures behave differently than a single tube. Their complexity arises mainly from the tube-tube interactions and the presence of van der Waals bonding. For example the thermal conductivity has been shown to drop from several thousand to 200 W/mK due to tube-tube interactions and tube bending and physical contacts[9]. Qi et al. conducted indentation tests on isolated aligned MWCNT samples grown by plasma enhanced chemical vapor deposition, and determined an effective axial modulus of a tube to be 0.9-1.23 TPa and an effective bending modulus to be 0.91-1.24 TPa[5]. In turf form Cao et al. studied the cyclic compression behavior of CNTs that were released from a substrate; these turfs were made of vertically aligned CNTs[10]. Their results showed that these films had a much higher compressibility and much higher strength than usual foam material. They showed that the stress-strain curve of the CNT film showed similar trends as foam materials; three stages occurred, with the first being a linear stress-strain relation with a high modulus (50 MPa), followed by a plateau characterizing the formation of zig-zag reversible buckles (at 22 % strain) and a reduction in modulus to 12 MPa, followed by a densification stage. This current paper will examine the deformation during compression of turfs still affixed to their growth substrate.

EXPERIMENT

Patterns of vertically aligned CNTs were selectively grown upon photolithographically prepared silicon wafers[8]. The catalyst was formed using a sol-gel solution with a base ethanol, TEOS, and $Fe(NO_3)_3$. The native oxide was removed from a polished silicon wafers, and the sol-gel was deposited on the Si wafer, which was then spun at 3000 rpm for 30 seconds and then dried for 24 hours at 80°C. The film was patterned using standard photolithography techniques. Carbon nanotube turfs were then grown using a previously described chemical vapor deposition technique[11].

Measurement of the tangent elastic modulus of the CNT turfs was performed using a Hysitron Triboscope in conjunction with a Park Scientific Autoprobe CP scanning probe microscope. A Berkovich tip with a tip radius of 1078 nm was used to conduct the experiments. After four indentations in the CNT turf, the diamond tip was cleaned to remove any residual CNTs stuck to the diamond tip due to strong adhesive bonding between the tip and CNTs. The indentation loading sequence included a loading segment, a hold segment, then an 80% unloading segment, a small holding segment from which drift is calculated, and a final unloading segment. The elastic modulus of the turf was determined using Oliver and Pharr's technique[12].

Two different techniques were used to buckle the CNT turfs. Several turfs with a 300 μm diameter and heights varying between 25 and 204 μm were tested using a displacement controlled compression tool that simultaneously recorded displacement with a LVDT and the load from a load cell attached to a spherical steel tip with a diameter of 1.6 mm. The spherical shape accommodates any misalignment between itself and the 300 μm diameter turf. Other turfs were tested in compression using a load controlled MTS Nanoindenter XP, with a flat diamond punch to contact the turf.

The CVD grown turfs were therefore tested under compression and permanently deformed in a localized manner using nanoindentation as shown schematically in figure 1a, and in a collective manner, shown in figure 1b.

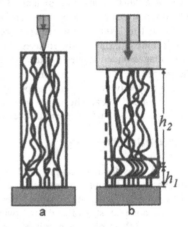

Figure 1 Schematic of deformation of CNT turfs under (a) nanoindentation and (b) flat punch compression.

RESULTS

Figure 2 shows a typical indentation conducted on a CNT turf. Initially, the turf softened under applied compressive loading, reflected by a convex curvature on the loading segment of the resulting load depth curve. However, the curve reaches an inflection point beyond which the turf ceased softening and loading is more linear (though still not reaching the concave curvature expected for a self similar indenter). The loading segment was then followed by hold segment which shows significant creep. The increased displacement during creep results in a lower applied load to the turf due to the load carried by the support springs. Creep in this system is

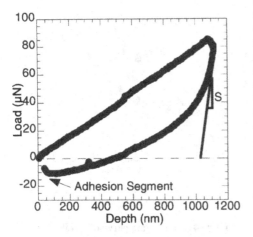

Figure 2 A typical nanoindentation load-depth curve of a CNT turf.

most likely due to tubes in contact that are thermally activated and sliding along each other. The unloading segment was characterized by a sharp initial slope which quickly decayed. As the load applied on the sample was decreased, the tip remained deflected downward due to adhesive forces between the tip and turf. These strong adhesive forces between the turf and the diamond tip result in a tensile load once the adhesive load surpasses that applied to the system, and the support springs apply a tensile load to the tip / turf system. Using a conventional unloading analysis, the turf's elastic modulus was 14.9 MPa ± 5.7 MPa and the hardness is on the order of 2 MPa.

The unloading segment reached a minimum (the maximum adhesive force), prior to a load increase at a constant slope (corresponding to the spring constant in the system) as shown in figure 2. During this process, the tip pulled CNTs upward, leaving no residual deformation at low loads. In order to determine the onset of plastic deformation of the turf under nanoindentation, the final portion of the unloading segment was fitted to a linear relationship, and the resulting intercept between the fit and the depth axis was determined. For an indent to be defined as completely reversible, the curve fit had to intercept the depth axis at its origin. This procedure was applied to several indents at increasing maximum applied loads, and the onset of plastic deformation was determined to be the maximum applied load of the indent that showed the first deviation from the depth axis origin. The onset of plastic deformation was 45 μN, corresponding to a compressive stress of 2.5 MPa.

Permanent deformation of a full turf, under applied compressive stress, was also addressed. A flat punch indentation was carried out using a MTS Nanoindenter XP on CNT turfs with a radius of approximately 30 μm and a height of approximately 60 μm, and a second set of tests with a compression test and a spherical indenter were carried out on a wide range of turf heights. The turf showed a unique coordinated buckling phenomena which occurred close to the fully constrained boundary condition as is shown in figure 3. The buckled region revealed the formation of several wrinkles, oriented unidirectionally. The load-displacement data collection were converted to stress-strain curves for the compression tests, as shown in figure 4.

The stress-strain curves showed a behavior similar to that of non linear buckling. Initially the stress increased with respect to strain until the buckling stress was reached. Following buckling, due to the displacement control in the test a slight decrease in applied stress was observed, followed by a stiffening segment as observed in figure 4 for both a flat punch and spherical tip of two different turfs. The difference between the buckling stress and the instability stress is likely related to the amount of energy required to cause a unidirectional reorientation of CNT segments within the localized buckling region as shown on the SEM image of Figure 3.

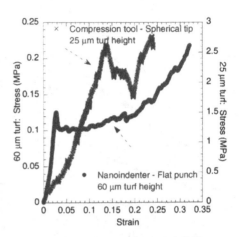

Figure 3 SEM image of a turf, following a compression test, showing localized, coordinated buckling, close to the rigid boundary condition.

Figure 4 Stress-Strain curve of CNT turf under applied compressive loading for a 60 and 25 μm high turf.

Based on the shape of the buckled turfs and knowing the elastic properties from nanoindentation experiments, a phenomenological model predicting the buckling stress of a turf was developed. At its buckled condition, the turf was divided into a buckling height h_1 defined from the rigid boundary condition to the center of the buckling wrinkle, and a top shear height, assumed to purely shear with no bending under applied loading, to accommodate for the lateral constraint enforced by the rigid flat punch in contact. The shape of a buckled CNT segment is given by

$$P(u_1 - u) - H(h_1 - z) = E_t I \frac{d^2 u}{dz^2} \tag{1}$$

where the load carried by the tubes is P and the lateral force, H resulting from the shear at height h_1 (see fig. 1),

$$H = \frac{\mu u_1}{h_2 N}, \tag{2}$$

depends on the effective shear modulus of the upper part of the turf, μ, the maximum lateral displacement u_1, and N, the density of the tubes (length per volume), E_t is the elastic modulus of a single tube, and I is the area moment of inertia of the turf's cross-sectional area. However, since we cannot measure E_t directly, we measure an effective modulus of the turf E, which will provide a continuum approximation that will account for the tube density and morphology. Solving this differential equation while employing a shear modulus $\mu = E/2$ and knowing that a CNT turf behaves like a foam material[10], and thus has a Poisson ratio $\nu = 0$[13], the following solution corresponding to a non-trivial solution $u_1(h_1) \neq 0$ (where u_1 is the maximum lateral displacement upon buckling) is derived:

$$\frac{\mu}{\sigma} = \frac{1 - h_1/h}{h_1/h} \qquad (3)$$

Several turfs with heights varying between 25 and 200 μm were then buckled using the compression tool. The buckling height of the turfs, h_1, was approximately 11 μm for most compression tests, independent of the height of the turf. The results of the prediction of Eq. (3) and the experimental result and the nanoindentation onset of permanent deformation stress are plotted on figure 5 against turf height, after being normalized by their respective elastic modulus.

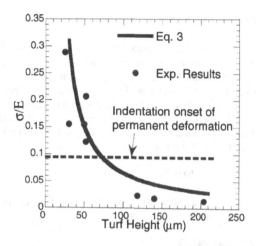

Figure 5 Buckling over E vs. turf height, along with the indentation onset of permanent deformation stress.

DISCUSSION

Based on the experimental tests in this study, one possible mechanism of permanent deformation in the turf is as follows. As the applied load increases, the tubes initially in contact were pushed further downward while sliding along adjacent CNT tubes and their morphology changed with respect to their initial morphologies. This bending, bowing, and sliding could be responsible for the softening behavior at the start of the indentation, as has been noted in earlier work[6,8]. Also, as the contact depth increased, more tubes will contact the tip, thus increasing the actual contact area. Once a critical load and depth is reached, the tubes can either become permanently entangled with each other, or can align under stresses to increase the contact area subjected to van der Waal bonding, thus causing the onset of permanent deformation. The onset of plastic deformation may be the case when the strain energy caused by bending and bowing of the tubes is exceeded by the contact energy due to an increase in the density of CNT segments contacting adjacent segments.

The major difference between the permanent deformation of a portion of a turf under nanoindentation and that of a buckling a turf is that under nanoindentation, the region being deformed is slightly constrained by the intact surrounding. In general, buckling occurs at stresses lower than that which caused permanent deformation in the indentation case. Since the turf is an assemblage of tubes that are entangled, there must be load transfer from tubes being compressed during indentation to tubes in the surrounding turf, and this additional region may well constrain the ability of the tubes under compression to freely deform, which increases the stress required for permanent deformation. In either the localized or uniform compression case, the contact loading response of a turf is complicated by adhesion and viscoelastic behavior that will be the subject of future studies.

CONCLUSIONS

Two types of contact loading tests were conducted on CNT turfs, a localized indentation test and a uniform compression test. The nominal elastic modulus and hardness, determined using an unloading slope analysis on indentations, were 14.9 MPa ± 5.7 MPa and 2 MPa respectively. A localized buckling phenomena has been observed in the case of uniform compression on a turf bonded to it's growth substrate. A phenomenological model predicting the buckling of CNT turfs was developed and verified experimentally. The differences between the buckling stress and the hardness and plastic deformation have been discussed in light of the constraint of the turf structure.

ACKNOWLEDGMENTS

The authors thank the National Science Foundation under grant CTS-0404370 for financial support of this project.

REFERENCES

1. Treacy, M. M. J., Ebbesen, T. W. & Gibson, J. M. *Nature* **381,** 678 (1996).
2. Baughman, R. H., Cui, Ch., Zakhidov, A., Iqbal, Z., Barisci, J., Spinks, G., Wallace, G., Mazzoldi, A., De Rossi, D., Rinzler, A., Jaschinski, O., Roth, S. & Kertesz, M. *Science* **284,** 1340 (1999).
3. Berber, S., Kwon, Y. amd Tomanek D. *Phys. Rev. Lett.* **84,** 4613 (2000).
4. Fan, Sh., Chapline, M. G., Franklin, N. R., Tombler, T. W., Cassell, A. M. & Dai, H. *Science* **283,** 512 (1999).
5. Qi, H. J., Teo, K.B.K., Lau, K.K.S., Boyce, M.C., Milne, W.I., Robertson, J. & Gleason, K.K. *J. Mech. Phys. Solids.* **51,** 2213 (2003)
6. Mesarovic, S. Dj., McCarter, C. M., Bahr, D. F., Radhakrishnan, H., Richards, R. F., Richards, C. D., McClain, D. & Jiao, J. *Scripta Materialia* **56,** 157 (2007).
7. Waters, J. F., Riester, L. Jouzi, M. Guduru, P. R. & Xu, J. M.. *Appl. Phys. lett.* **85,** 1787 (2004).
8 . McCarter, C. M., Richards, R. F., Mesarovic, S. Dj., Richards, C. D., Bahr, D. F., McClain, D. & Jiao, J. *J. Mater. Sc.* **21,** 7872 (2006).
9. Yang, D. J., Wang, S. G., Zhang, Q., Sellin, P. J. & Chen, G. *Phys. Lett. A.* **329,** 207 (2004).
10. Cao, A. Dickrell, P. L., Sawyer, W. G., Ghasemi-Nejhad, M. N. & Ajayan, P. M. *Science* **310,** 1307 (2005).
11. Dong, L., Jiao, J., Pan C. and Tuggle, D. W. *Appl. Phys. A* **78,** 9-14 (2004)
12. Oliver, W. C. & Pharr G. M. *J. Mater. Res.* **7,** 1564-1583 (1992).
13. Deshpande V. S. and Fleck, N. A., *J. Mech. Phys. Solids* **48,** 1253 (2000).

Mater. Res. Soc. Symp. Proc. Vol. 1049 © 2008 Materials Research Society

Effect of Hydrostatic Pressure on Indentation Modulus

W. M. Mook, and W. W. Gerberich
University of Minnesota, Minneapolis, MN, 55455

ABSTRACT

The high pressures generated at a contact during nanoindentation have a quantifiable effect on the measured indentation modulus. This effect can be accounted for by invoking a Murnaghan equation of state-based analysis where the measured indentation modulus is a function of the hydrostatic component of the stress state which is generated beneath the indenter tip. This approach has implications pertinent to a range of mechanical characterization techniques that include instrumented indentation and quantitative atomic force microscopy (AFM) since these techniques traditionally consider only zero-pressure modulus values during data interpretation. To demonstrate the validity of this approach, the indentation modulus of four materials (fused quartz, sapphire, rutile and silicon) is evaluated using a 1 μm radius conospherical diamond tip to maximum contact depths of 30 nm. The tip area function is independently determined via AFM while the unloading stiffness from the load-displacement data is determined using standard Oliver-Pharr analysis.

INTRODUCTION

Nanoindentation has become one of the most widely used characterization techniques for quantifying the mechanical properties of thin films and freestanding structures [1, 2] at length scales below 1 μm. The method most often employed for nanoindentation data analysis is that of Oliver and Pharr (OP) [3]. This technique assumes that the initial point of unloading is purely elastic. Thus the initial slope of the load-displacement data is the elastic stiffness, S, and is defined by

$$S = \beta \frac{2}{\sqrt{\pi}} E_r \sqrt{A} \qquad (1)$$

where β is a dimensionless parameter, A is the projected area of contact and E_r is the reduced modulus of the tip-substrate system. The reduced modulus is

$$\frac{1}{E_r} = \frac{1-v_1^2}{E_1} + \frac{1-v_2^2}{E_2} \qquad (2)$$

where v is Poisson's ratio and the subscripts refer to the tip and the substrate respectively. The other common parameter typically measured using nanoindentation is the material's hardness, or average contact stress, $\bar{\sigma}_c$, which is

$$\bar{\sigma}_c = \frac{P_{max}}{A}, \qquad (3)$$

where P_{max} is the maximum load.

A recent review of the OP method [4] addresses modifications to the initial theory. One of the major findings has been the variability of β which has now been shown to depend on a number of physical parameters. These include tip shape, tip geometry and tip-sample friction among others [5]. While refinements have been made, further analytical improvements are still warranted considering the results of the recent NIST nanoindentation round robin where a single sample (1.5 μm thick Cu film deposited on a Si substrate) was diced and distributed to a number of independent labs [6]. The variations in reported hardness and modulus values between the labs were larger than what was expected with one standard deviation for hardness of 15% and modulus of 19%. It seems likely that unaccounted for differences in the β value of Eq. (1) could contribute to the observed spread in reported values. Considering the many physical parameters that affect β, we would like to propose that this dimensionless parameter may be, in part, a physical manifestation of the hydrostatic component of stress that is created beneath the indenter tip during indentation.

While the effect of hydrostatic pressure on indentation modulus has been largely ignored, geophysicists have been quantifying its effect on bulk modulus for decades by using seismic events to investigate the properties of the earth's high-pressure interior. This has led to the development of equations of state applicable to solids. One of the most simple, yet most popular and effective equations of state is the Murnaghan relationship [7, 8]. A basic assumption that can be used to derive this relation is that the bulk modulus is a linear function of hydrostatic pressure for small volume changes, such that

$$K = K_0 + K_0'\sigma_H, \tag{4}$$

where K is the bulk modulus, K_0 is the zero pressure bulk modulus, K_0' is the bulk modulus pressure first derivative and σ_H is hydrostatic pressure. It will be shown that this, albeit in a slightly modified form, can be applied to indentation analysis. This relation has been confirmed using an ever-growing database of experimental seismic [9] and diamond anvil cell (DAC) [10, 11] data. A value of $K_0' \sim 4$ is found to be valid for many ceramic materials [7, 12].

Only recently has the effect of hydrostatic pressure on indentation modulus been considered. Finite element simulations have shown that if one accounts for pressure, indentation modulus is increased [13] and that the component of hydrostatic pressure can be equivalent to one-half of the average contact stress for a spherical tip indenting a half-space [14]. Experimentally, it has been shown that when subjected to uniaxial compression, pressure increases the modulus of single crystalline nanoparticles [2]. However an experimental assessment accounting for hydrostatic pressure effects on a traditional indentation geometry has yet to be undertaken. To do this, a material's bulk modulus can be related to its elastic modulus, E, through Poisson's ratio, v, by

$$K = \frac{E}{3(1-2v)}. \tag{5}$$

By combining Eqs. (4) and (5), one finds,

$$E = E_0 + M\sigma_H, \tag{6}$$

where M is a material-specific constant which is composed of K_0' and v such that

$$M = 3K_0'(1-2v). \tag{7}$$

While v can be affected by hydrostatic pressure, we are treating it as a constant for the pressure range under consideration. In this range for crystalline materials, v should slightly decrease with pressure which implies M should increase. This slight increase in M is accounted for in the R function described below. Equation (6) describes elastic modulus as a function of hydrostatic pressure. For materials with $v < 0.33$, Eq. (6) predicts that E is more sensitive to σ_H than K. Of course using standard OP analysis, it is only possible to measure the hardness (average contact stress) and not the hydrostatic component of stress. Therefore an effective stress ratio, $R = \sigma_H / \bar{\sigma}_c$, is employed in order to determine the hydrostatic component of the stress state as a ratio of the contact stress. Elastic modulus as a function of contact stress can be determined as

$$E = E_0 + MR\bar{\sigma}_c,\qquad(8)$$

where Eq. (8) is applied to both the tip and substrate to generate the indentation modulus from Eq. (2).

To demonstrate the validity of this approach, the indentation modulus of four materials (fused quartz, sapphire, rutile and silicon) is evaluated using a 1 μm radius conospherical diamond tip to maximum contact depths of 30 nm. The tip area function is independently determined via an AFM-based technique while the unloading stiffness and contact displacement is determined from the load-displacement data using standard Oliver-Pharr analysis. Values for E_r, $\bar{\sigma}_c$, β and R as a function of contact displacement are calculated and discussed.

EXPERIMENT

In order to create an independent area function, the 1 μm radius tip is imaged by scanning it over a grating of Si spikes (TGT1 from NT-MDT) using a Hysitron TriboScope® that is attached to a DI NanoScope® III AFM. The spikes have a quoted radius of curvature of 10 nm. A 3×3 μm image is generated at 256×256 resolution thereby imparting lateral dimensions for each pixel of approximately 11.7 nm. A routine was generated to slice across the tip+spike image and to directly count the pixels. This generates a graph of the projected area as a function of the first 30 nm of displacement from the apex of the tip seen in Figure 1(a) and is fitted with a two-term area function where δ is displacement

$$A_{AFM} = 42.346(\delta^2) + 6419.3(\delta); \quad \left[R^2 = 0.9997 \right].\qquad(9)$$

For reference, the ideal area function for a sphere ($R=1$ μm) is also graphed in Figure 1(a) and is

$$A_{ideal} = -\pi(\delta^2) + 2\pi R(\delta).\qquad(10)$$

The tip+spike image overestimates the true area of the tip which will generate a lower bound for both indentation modulus and average contact stress throughout this displacement range.

Indentations were conducted under load-control conditions using a Hysitron TriboIndenter®. A machine compliance of 5.1 nm/mN was applied to all data and was calibrated with a standard algorithm [15] using fused quartz. All samples were mounted with Crystalbond™ and surfaces were measured to be normal to the loading axis within 2% with an RMS surface roughness of <1 nm as determined by contact-mode AFM. Prior to indentation, the tip was brought into contact with the sample and allowed to scan the surface with a scan size of 5 μm at a rate of 0.5 Hz with a normal load of approximately 1 μN. Scanning continued until the drift rate decreased below 0.1 nm/s. The tip is then raised 5 nm above the sample at which point the load-control indent is initiated at a constant rate not greater than 500 μN/s. The tip is held at the maximum load for five seconds prior to unloading which is conducted at the same rate as the loading segment. After each indent, the tip is moved 5-10 μm to the next location. The materials, which included (0001) Al$_2$O$_3$ sapphire, (001) Si, (001) TiO$_2$ rutile and fused quartz, were chosen since they can be polished to very smooth surfaces and are stiff enough to sink-in under elastic loading. These conditions are required in order to accurately determine the contact displacement, δ_c, with OP analysis [4]. The load-displacement data for the indents can be seen in Figure 1(b).

The relevant material properties used in the analysis below can be found in Table I.

Table I. Material Properties: [a][8, 12], [b]in the range of [16, 17], $E_{0(r)}$ is Eq. (2), M is Eq. (7).

Material	ν	E_0 (GPa)	$E_{0(r)}$ (GPa)	K'_0 (GPa)	M
Diamond	0.07	1000	-	3.5 [a]	9.1
Fused Quartz	0.17	72	69.9	4.5 [a]	8.9
Sapphire	0.25	380	288.9	4.5 [a]	6.8
Rutile	0.28	277	231.3	9 [b]	11.9
Silicon	0.21	160	143.5	4 [a]	7.0

DISCUSSION

Examination of Figure 1(b) shows that the loading for each material was nearly elastic with residual displacements ranging from 0-2 nm. The indentation modulus seen in Figure 2(a) is calculated from Eq. (1) assuming, for now, that β is unity as it is predicted to be for a spherical indentation [4]. The modulus values all tend to increase with decreasing displacement. However the average contact stress depicted in Figure 2(b), as calculated from Eq. (3), shows the opposite trend. This is because the indents are still mainly elastic; indentation to further displacements generates plasticity.

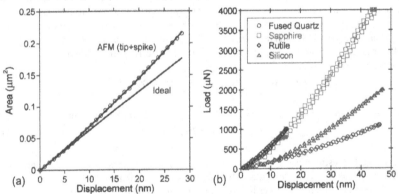

Figure 1: (a) AFM-based area function of the 1 μm radius tip created by scanning it over a grating of Si spikes. The tip+spike area function is seen in Eq. (9) while Eq.(10) is the ideal spherical function. (b) Representative load-displacement data for indentation into sapphire, silicon, fused quartz and rutile by a 1 μm radius conospherical diamond tip.

From Figure 2(a), it is possible to calculate an effective β_{eff} value which is the measured indentation modulus divided by the zero pressure indentation modulus, the results of which can be seen in Figure 3(a). It is apparent that while the β_{eff} values for sapphire, silicon and fused quartz follow the same trend, the values for rutile deviate sharply. In other words β_{eff} is different for different materials. The data also indicates that β_{eff} is a function of displacement as well, but that may be due to the upper bound estimate of the tip+spike area function which, as seen in Figure 1(a), tends to deviate from an ideal sphere as displacement increases.

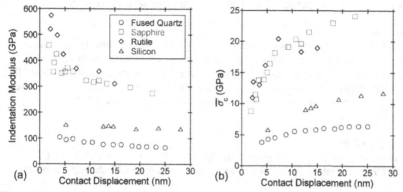

Figure 2: Analysis of both indentation modulus (Eq. (1)) and average contact stress (Eq. (3)) from the load-displacement data from Figure 1(b) using the area function generated in Figure 1(a).

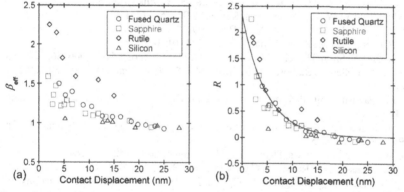

Figure 3: Quantification of the increase in indentation data seen in Figure 2 where (a) β_{eff} is the ratio of measured modulus to zero-pressure modulus, while in part (b) R is the effective stress ratio for this tip-contact area combination. The data from Figure 2 is analyzed using Eq. (8) which is applied to both tip and substrate to generate R as a function of displacement.

From the data in Figure 2 it is also possible to calculate the effective stress ratio, R, that is found in Eq. (8), the results of which can be seen in Figure 3(b). The data, including rutile, collapses onto the same trend line. This is because rutile is quite sensitive to hydrostatic stress as can be seen in Table I where its K_0' is up to twice the value of the other materials. In other words when under contact, rutile's bulk and elastic moduli increase dramatically under pressure which explains its nanoscale load-displacement response. As seen in Figure 1(b), both rutile and sapphire initially have the same stiffness values and displacements. It should be noted that rutile can transform to various high pressure phases, all of which increase in density [18]. We are not likely observing that here since transformation would lead to a pop-in during the loading similar to results seen in silicon. Pop-ins are only observed for deeper indents into rutile using this 1 μm tip. Future work will address creating a more robust area function which should limit $R < 1$.

CONCLUSIONS

Traditional data analysis of indentation results employs zero-pressure elastic modulus values. However in view of the material specific constant, M, for materials with $v < 0.33$ hydrostatic pressure should influence elastic modulus more than bulk modulus. It is believed that by accounting for hydrostatic pressure effects, the accuracy of indentation data analysis can be improved. Therefore we have presented a methodology that attempts to account for this dependence to first-order.

ACKNOWLEDGMENTS

This work funded through NSF grant CMS-0322436 and was carried out at the U. of Minnesota I.T. Characterization Facility which receives support from NSF through the NNIN.

REFERENCES

1. W.W. Gerberich, W.M. Mook, M.J. Cordill, J.M. Jungk, B. Boyce, T. Friedmann, N.R. Moody and D. Yang, Int. J. Fracture, 2006, 138, 75-100.
2. W.M. Mook, J.D. Nowak, C.R. Perrey, C.B. Carter, R. Mukherjee, S.L. Girshick, P.H. McMurry and W.W. Gerberich, Phys. Rev. B, 2007, 75, 214112-214111.
3. W.C. Oliver and G.M. Pharr, J. Mater. Res., 1992, 7, 1564-1583.
4. W.C. Oliver and G.M. Pharr, J. Mater. Res., 2004, 19, 3-20.
5. J.H. Strader, S. Shim, H. Bei, W.C. Oliver, G.M. Pharr, Philos. Mag., 2006, 86, 5285-98.
6. D. Read, R. Keller, N. Barbosa and R. Geiss, Metall. Mater. Trans. A, 2007, 38, 2242-2248.
7. F.D. Murnaghan, Finite deformation of an elastic solid, John Wiley and Sons, Chapman and Hall, 1951.
8. O.L. Anderson, Equations of State for Solids in Geophysics and Ceramic Science, Oxford U. Press, Oxford, 1995.
9. F.D. Stacey and P.M. Davis, Phys. Earth Planet. Inter., 2004, 142, 137-184.
10. N.E. Christensen, A.L. Ruoff and C.O. Rodriguez, Phys. Rev. B, 1995, 52, 9121-9124.
11. S. Speziale, Z. Chang-Sheng, T.S. Duffy, R.J. Hemley and M. Ho-Kwang, J. Geophys. Res., 2001, 106, 515-528.
12. J.P. Poirier, Introduction to the physics of the Earth's interior, 2nd ed., Cambridge University Press, Cambridge, New York, 2000.
13. R.G. Veprek, D.M. Parks, A.S. Argon and S. Veprek, Mater. Sci. Eng., A, 2007, 448, 366-378.
14. B. Wolf and M. Goken, Z. Metallkd., 2005, 96, 1247-1251.
15. A.C. Fischer-Cripps, Nanoindentation, New York : Springer, 2004.
16. L. Gerward and J.S. Olsen, Mater. Sci. Forum, 1996, 228-231, 383-386.
17. R.M. Hazen and L.W. Finger, J. Phys. Chem. Solids, 1981, 42, 143-151.
18. L. Gerward and J.S. Olsen, J. Appl. Crystallogr., 1997, 30, 259.

Size Effects and Indentation
of Thin Films

Mater. Res. Soc. Symp. Proc. Vol. 1049 © 2008 Materials Research Society 1049-AA03-03

Microstructural Investigation of the Deformation Zone below Nano-Indents in Copper

Martin Rester, Christian Motz, and Reinhard Pippan
Erich Schmid Institute, Austrian Academy of Sciences, Jahnstrasse 12, Leoben, 8700, Austria

ABSTRACT

The deformation zone below nanoindents in copper single crystals with an $< 1\bar{1}0 > \{111\}$ orientation is investigated. Using a focused ion beam (FIB) system, cross-sections through the center of the indents were fabricated and subsequently analyzed by means of electron backscatter diffraction (EBSD) technique. Additionally, cross-sectional TEM foils were prepared and examined. Due to changes in the crystal orientation around and beneath the indentations, the plastically deformed zone can be visualized and related to the measured hardness values. Furthermore, the hardness data were analyzed using the Nix-Gao model where a linear relationship was found for H^2 vs. $1/h_c$, but with different slopes for large and shallow indentations. The measured orientation maps indicate that this behavior is presumably caused by a change in the deformation mechanism. On the basis of possible dislocation arrangements, two models are suggested and compared to the experimental findings. The model presented for large imprints is based on dislocation pile-ups similar to the Hall-Petch effect, while the model for shallow indentations uses far-reaching dislocation loops to accommodate the shape change of the imprint.

INTRODUCTION

It is well known for many years that the hardness of metals and alloys in the micron and sub-micron regime is not a constant number. In fact the hardness depends on the size of the indent i.e., with decreasing indentation depth the hardness increases [1-3]. This is called the indentation size effect (ISE). Using the concept of geometrically necessary dislocations (GNDs) and the Taylor rule for the flow stress, Nix and Gao (N-G) proposed a model to explain the ISE [4]. According to this model, a linear correlation between H^2, the square of the hardness, and $1/h_c$, the reciprocal indentation depth exists, which is in good agreement with micro-indentation hardness data. In the literature, however, it is reported that nanoindentation hardness data do not follow this linear trend over the whole measurement range [5-7]. Instead, at small indentation depths they start to deviate from the predicted linear curve. In order to verify if this behavior is linked to a change in the deformation structure the plastically deformed volume below different sized nanoindentations is visualized using electron backscatter diffraction (EBSD) and transmission electron microscopy (TEM). The results were subsequently used to suggest possible dislocation arrangements in order to explain the indentation process of large and shallow imprints.

EXPERIMENT

Copper single crystals with an $< 1\bar{1}0 > \{111\}$ orientation were prepared by wet grinding and mechanical polishing. Electropolishing was subsequently performed on the $\{111\}$ surface in order to remove any deformation layer produced during previous polishing steps. Using a Hysitron Triboscope fitted with a cube corner indenter the hardness and the indentation modulus of the material at loads between 40 μN and 10 mN were determined. A calibrated area function was obtained using a procedure outlined by Oliver and Pharr [8]. For all indentations the load function described in [9] was used. The presented results are an average of three to five indentations of each selected indenter load.

Cross-sections through the imprints were produced using a focused ion beam (FIB) workstation. For this purpose, several indentations with loads between 0.5 mN to 10 mN were produced in the vicinity of a carefully mechanically polished edge. To protect the imprints against Ga$^+$ ion damage a 500 nm thick tungsten layer was deposited over the indents. Prior to depositing, the centre of the indents were marked in order to get the cross-sections right through the middle of the indents. Using milling currents of 10 nA, 500 pA and 200 pA, respectively, the material in front of the indents was removed. On the readily polished cross-sections EBSD investigations were performed using a Leo 1525 field emission scanning electron microscope (SEM) equipped with an EDAX EBSD system. Since plastic deformation is usually associated with crystal orientation changes, traces of the plastic deformation can be visualized.

TEM samples were also prepared using the FIB workstation. Indents with loads between 500 μN and 10 mN were made and protection layers were deposited. By cutting two trenches on each side of the indents, a foil with a thickness of approximately 2 μm was fabricated. After lifting out, the foil was thinned to electron transparency using acceleration voltages of 30 kV and 5 kV, respectively. All TEM examinations were performed using a Philips CM12 TEM operating at 120 kV.

DISCUSSION

The results of the hardness measurement obtained for the {111} surface of copper are presented in Figure 1. As can be seen, the hardness data show a pronounced ISE. Starting at a value of 2.75 GPa the hardness decreases to 1.1 GPa at a depth of 1.8 μm. The reduced indentation modulus is over the whole measuring range on a rather constant level of approximately 125 GPa.

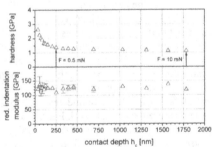

Figure 1. Hardness and reduced indentation modulus as a function of the contact depth. The hardness values of the largest and the smallest microstructurally investigated imprints are marked by arrows.

In order to verify if the Nix-Gao model fits the hardness data, the square of the hardness is plotted against the reciprocal indentation depth. The resulting plot is displayed in Figure 2, with an inset showing the hardness data for depths greater 100 nm enlarged. What can be observed is a hardness curve which consists of two linear regimes. Regime α thereby extends from the largest indentation depth to a depth of about 1 μm, whereas regime β describes the hardness for imprints smaller than 1 μm. This is somewhat different to the results found in literature, where the hardness data at small indentation depths starts to deviate in a non-linear way [5-7]. To find an explanation for the observed bi-linear behavior, the microstructure of the deformation-induced zone below nanoindentations was investigated by means of EBSD and TEM technique. The results of the EBSD and TEM studies for a 10 mN and a 0.5 mN indent are shown in Figure 3 (a - d). Starting with the misorientation map of the large indent (a), on the left hand side two deformation-induced rotation patterns (I and II), which are separated by a boundary, can be found.

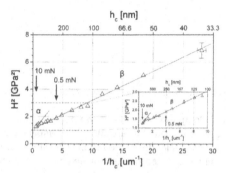

Figure 2. Application of the Nix-Gao model to the measured hardness data. The inset shows the hardness data for depths greater than 100 nm enlarged. The arrows mark the largest and the smallest of the microstructurally examined indentations.

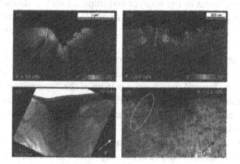

Figure 3. Misorientation maps (a - b) and TEM micrographs (c - d) for a 10 mN and 0.5 mN indent in copper {111} single crystals. The TEM micrographs were taken with a $\bar{1}1\bar{1}$ two beam condition.

Another interesting part is the region below the indenter tip, where a kind of sub-grain formation occurs. Both characteristic features can be found as well in the TEM micrograph as illustrated in Figure 3 (c). However, the deformation-induced zone of the small imprint looks somewhat different. A pronounced boundary as found for the 10 mN indent can not be observed. Rather, pattern I and II are separated only by a less defined boundary (Figure 3 (b)). This fact is confirmed by the TEM micrograph where instead of a sharp boundary single dislocations appear (the region is marked by the white ellipse in Figure 3 (d)). Both the bi-linear behavior of the hardness data (Figure 2) as well as the results obtained from the EBSD and TEM studies indicate a change in the deformation mechanism. In order to explain the observed behavior, possible arrangements of GNDs reflecting the characteristic features are presented in Figure 4.

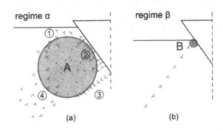

Figure 4. Dislocation models describing the indentation process of large (a) and shallow (b) imprints [9].

For large indentations a pile-up model similar to those used to describe the Hall-Petch relation is suggested. Since large indentations are always accompanied by the occurrence of a huge and far reaching shear stress field, dislocation sources located near the indenter flank (region A in Figure 4 (a)) can be activated and are able to emit dislocation loops. To simplify matters, only two types of slip planes, one perpendicular to the indenter flank and the other parallel to it, are assumed to generate dislocations. Dislocation loops emitted on slip planes perpendicular to the indenter flank start to move towards the indenter and pile up in front of the indenter, producing the required large orientation changes (region 2). This results in a pile-up inducing a significant back stress to the sources and impedes further dislocation generation. Dislocations with an opposite sign move into the opposite direction towards the bulk material (region 4) and arrange there in a very widespread manner. Consequently, the induced orientation gradient is only small.
Dislocation loops generated on the slip planes parallel to the indenter flank move into the region below the indenter tip. However, due to a change in the shear stress field they are not able to overcome the centre region. Instead they form a pile up at the "symmetry" line (region 3). The other parts of the loops move towards the free surface where few of them exit the material. But the majority of the dislocations arrange themselves in the region where the shear stress goes to zero forming the observed boundary (region 1).

But why does the described mechanism become less important when the indentation depth is decreased? Lowering the indentation depth is directly linked to a diminishment of region A and consequently to a decrease of the number of dislocation sources which can be activated. This results in an increase of the back stress originating from the dislocations piled up in region 2 and 3. As a result, the generation of further dislocations is impeded. Due to the hindered dislocation emission other mechanisms like dislocation generation laterally of the indent becomes more important. The emitted dislocations form a kind of prismatic loops which move on slip planes arranged very close to each other. As a consequence, the recent generated dislocations have to push the previously created ones towards the bulk material. However, for extremely shallow indents the segment length of the dislocations generated in region B becomes very small and therefore the stress required to push the dislocations away from the indenter flank is very high.

Increasing the indentation depth and thus the segment length of the dislocations causes the observed decrease of hardness.

To verify if the suggested models are realistic, the required shear stress to obtain such a dislocation arrangement is estimated. To convert the received stress into hardness, Tabor's rule was used [10]. For large indentations the pile-up model described in Figure 4 (a) seems to be responsible for the ISE. Since there are similarities to the Hall-Petch effect (H-P), the H-P relation

$$\tau_\alpha = \tau_{HP} = \tau_0 + \frac{k'_{HP}}{\sqrt{h_c}} \qquad (1)$$

is used to calculate the shear stress for regime α [11,12]. The grain size thereby is substituted by the indentation depth h_c which is in turn proportional to the size of region A and τ_0 and k'_{HP} are constants. However, since the kink in the N-G plot (Figure 2) refers to a change in the deformation mechanism this type of stress calculation is only valid for regime α. For regime β, the stress is estimated in a different way. As can be seen in the dislocation model presented for shallow nanoindentations (Figure 4 (b)) single dislocation events become more and more important. Subsequently, the size of dislocation sources as well as the back stress originating from previous emitted dislocations has to be considered for stress calculation as well [13-15]. Thus, the following relation can be found

$$\tau_\beta = \tau_{source} + \tau_{back} = \frac{Gb}{2\pi s}\ln(\frac{\alpha s}{b}) + \frac{Gb}{2\pi^2 D}\ln(\frac{\sqrt{2}h_c}{b}) \qquad (2)$$

Here $G = 47$ GPa is the polycrystalline shear modulus of copper, $b = 0.256$ nm is the Burgers vector, s the source size which was assumed to be $2h_c$ and α a numerical constant in the order of unity. D, the diameter of the area where dislocations were emitted is assumed to be 150 nm. Converting the calculated stress into hardness values and plotting them in the N-G plot shows good agreement (Figure 5).

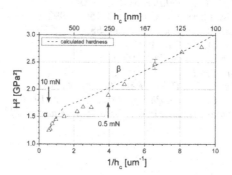

Figure 5. Comparison of the measured hardness data to the calculated hardness for depths up to 100 nm using a Nix-Gao plot. The arrows mark the largest and the smallest of the examined indentations.

Special attention was paid to the fact that the used values for s, the source size and D, the diameter of the region where dislocation emission takes place are plausible. Even though the performed calculations are only rough estimations, they nevertheless demonstrate that the supposed dislocation models are able to describe the observed bi-linear behavior of the hardness data.

CONCLUSIONS

Studying the N-G plot of the hardness data measured on {111} copper single crystals reveals two linear regimes, one for indents with depths greater 1 μm and one for indents smaller than 1 μm. It is assumed that this bi-linearity is linked to a change in the deformation mechanism. EBSD and TEM studies, performed for a large and a shallow indent support this assumption. On the basis of the experimental findings two models of possible dislocation arrangements are presented and used to explain the ISE. Subsequently, the shear stress field of the dislocation arrangements was estimated and by means of Tabor's rule compared to the measured hardness data. The hardness values of the simple estimation show good agreement with the experimental findings.

ACKNOWLEDGMENTS

This work was financially supported by the FWF (Fonds zur Förderung der wissenschaftlichen Forschung) through Project P 17375-N07.

REFERENCES

1. N. Gane and J. M. Cox, *Philos. Mag.* **22**, 881 (1970).
2. J. B. Pethica, R. Hutchings and W. C. Oliver, *Philos. Mag. A* **48**, 593 (1983).
3. Q. Ma and D. R. Clark, *J. Mater. Res.* **10**, 853 (1995).
4. W. D. Nix and H. Gao, *J. Mech. Phys. Solids* **46**, 411 (1998).
5. Y. Y. Lim and M. M. Chaudhri, *Philos. Mag. A* **79**, 2979 (1999).
6. J. G. Swadener, E. P. George and G. M. Pharr, *J. Mech. Phys. Solids* **50**, 681 (2002).
7. G. Feng and W. D. Nix, *Scripta Mater.* **51**, 599 (2004).
8. W. C. Oliver and G. M. Pharr, *J. Mater. Res.* **7**, 1564 (1992).
9. M. Rester, C. Motz and R. Pippan, *Acta Mater.* **55**, 6427 (2007).
10. D. Tabor, *Proc. R. Soc. A* **192**, 247 (1947).
11. E. O. Hall, *Proc. Phys. Soc. B* **64**, 747 (1951).
12. N. J. Petch, *J. Iron Steel Inst.* **174**, 25 (1953).
13. J. C. Fisher, E. W. Hart and R. H. Pry, *Phys. Rev.* **87**, 958 (1952).
14. L. H. Friedman and D. C. Chrzan, *Philos. Mag. A* **77** 1185 (1998).
15. B. von Blanckenhagen, E. Arzt and P. Gumbsch, *Acta Mater.* **52** 773 (2004).

Mater. Res. Soc. Symp. Proc. Vol. 1049 © 2008 Materials Research Society   1049-AA03-05

Microstructure and Mechanical Properties Characterisation of Nanocrystalline Copper Films

Nursiani Indah Tjahyono, and Yu Lung Chiu
Department of Chemical and Materials Engineering, University of Auckland, 20
Symonds Street, Auckland, 1142, New Zealand

ABSTRACT

The microstructure and mechanical properties of nanocrystalline copper with grain size ranging from 50 nm to 80 nm have been investigated. Nanocrystalline copper films were electrodeposited from an additive-free acidified copper sulphate solution at room temperature by employing constant current at different current density magnitudes between 20 and 80 mA/cm^2. Both austenitic and ferritic steel substrates with the same surface finishing conditions have been used for the deposition. The microstructure of the films has been further studied using electron microscopy techniques, and the mechanical properties using nanoindentation technique. The nanoindentation study was carried out on both the plan view and cross-sectional directions to study the isotropy characteristic of the copper film. It has been noted that both the modulus and hardness measured following the Oliver-Pharr scheme show an apparent indentation size effect tested on the cross-sectional sample.

INTRODUCTION

Over the years, there has been an increasing interest in the development of thin film systems. Understanding of the mechanical properties and deformation mechanism of these materials is crucial to the development of reliable devices. Nanoindentation technique has been developed to provide an easier approach to evaluate the mechanical properties of thin films without complicated sample preparation and significantly reduced the sample size requested for testing and is considered as a non-destructive method [1].

Thin films often contain very small sized grains (e.g. nanocrystalline films) and many studies have shown that they exhibit very different microstructure and mechanical behaviours in comparison to their conventional coarse-grained counterparts [2]. This is because nanocrystalline materials contain a relatively high grain boundary volume fraction which in turn influenced their mechanical properties, such as the yield strength, hardness and ductility, considerably [2, 3]. Furthermore, the film/substrate interface in the thin film system plays a role in restraining the dislocation motion thus increases the film hardness with decreasing film thickness [4].

Thin film copper has been widely used as an interconnect material in microelectronic devices as well as in the integrated circuit manufacturing industry [2, 5]. Many studies have been conducted to characterise the mechanical properties of the copper thin film including the effect of different substrates by Fang and Chang [6] and Beegan *et al* [7], the effect of copper film thickness by Volinsky *et al* [8] and the tribological properties of nanocrystalline copper by Tao and Li [5]. Nevertheless, most of these studies have focused on the plan view samples and studies on the cross section samples are rare, despite the fact that the mechanical properties are equally important on cross sections [9]. In this study, nanoindentation experiments have been carried out to compare the indentation hardness and modulus obtained from both the cross-sectional and plan view directions on the ferritic steel substrate, and from the cross-sectional direction in the austenitic steel.

EXPERIMENT

Nanocrystalline copper films were prepared using electrodeposition technique with constant current in an additive-free acidified copper sulphate electrolyte at room temperature. The ferritic substrate and austenitic steel substrates were mechanically polished down to 1 μm finishing and cleaned thoroughly prior to the deposition. The grain size of the nanocrystalline copper was controlled by changing the applied current density in the range from 20 to 80 mA/cm^2.

The mechanical properties of electrodeposited copper films were evaluated on a MTS Nanoindentation XP system with a sharp Berkovich tip, on both the plan view and cross sections. All samples were mechanically polished down to 0.04 μm before nanoindentation tests were carried out at a constant strain rate of 0.01 s^{-1}. The maximum indentation depth was controlled to be 1500 nm which accounted for less than 10% of the film thickness for all of the samples, except for the copper film deposited at current density of 20 mA/cm^2 on austenitic steel, where the film thickness was measured to be 16 μm and the maximum indentation depth was reduced to be 1100 nm.

COPPER FILM MICROSTRUCTURE

The sample designation and the grain sizes of the deposited copper films measured using Scherrer Equation based on the broadening of the X-ray diffraction peaks are shown in Table 1. It can be seen that with higher the applied current density, smaller grain size deposit can be obtained. At low deposition current density, the copper films deposited on the ferritic steel have quite similar grain size in comparison to those deposited onto the austenitic sample. At higher current density, however, the deposits on the ferritic steel substrate have slightly smaller grain size than those on the austenitic steel substrate. It has been found that the higher applied current density resulted in thicker film in the same deposition time. The thickness of the deposits ranges from 16 (AS1) to 80 μm (FS4).

Table 1. The grain size and thickness of the deposited copper films.

Current Density (mA/cm^2)	Ferritic Steel Substrate			Austenitic Steel Substrate		
	Designation	Grain size (nm)	Thickness (μm)	Designation	Grain Size (nm)	Thickness (μm)
20	FS1	76.3 ± 9.3	29.1 ± 1.4	AS1	75.2 ± 9.9	16.3 ± 0.8
40	FS2	69.0 ± 9.2	54.6 ± 3.0	AS2	69.7 ± 8.9	37.8 ± 2.7
60	FS3	60.5 ± 7.2	71.8 ± 5.6	AS3	65.4 ± 6.4	61.7 ± 4.1
80	FS4	53.8 ± 9.2	87.9 ± 5.6	AS4	57.0 ± 7.1	78.3 ± 6.7

Figures 1 and 2 show the surface morphology of the as-deposited copper films at 20 and 80 mA/cm^2 on the ferritic steel and austenitic steel substrates, respectively. It can be seen that the higher current density results in the higher surface roughness.

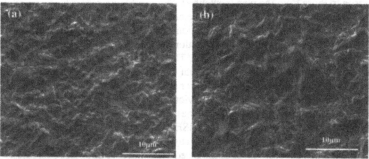

Figure 1. SEM micrographs of the as-deposited surfaces of sample FS1 (a) and FS4 (b).

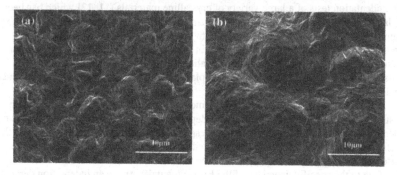

Figure 2. SEM micrographs of the as-deposited surfaces of sample AS1 (a) and AS4(b).

COPPER FILM MECHANICAL PROPERTIES

The hardness and elastic modulus of the electrodeposited copper film deposited onto ferritic steel substrate (i.e. sample F1-FS4) have been measured from the plan view and cross-sectional directions and are shown in Figures 3 and 4, respectively. In both cases, the hardness measured (Fig. 3b and 4b) decreases with increasing indentation depth, showing an indentation size effect (ISE). Chen and Vlassak [10] have found that, for the case of soft film on hard substrate, the intrinsic film hardness can be obtained by controlling the indentation depth less than 50% of the total film thickness. In the present study, the steel substrates is much harder than the copper films and furthermore the indentation depth is kept to less than 10% of the thickness of the film, thus, the substrate/film interface and the steel substrates have minimal contribution to the hardness measured. In other words, the hardness measured represents the intrinsic film response to the local penetration. The hardness measured on cross-sectional samples (Fig. 4b) has a slightly more pronounced ISE than that measured on plan view samples (Fig. 3b). Nonetheless, the hardness values of all four samples (FS1-4) are fairly constant. Furthermore, slightly higher hardness is obtained from the cross-sectional direction in comparison to those from the plan view direction.

The elastic modulus measured on the plan view samples shows fairly constant value of 140 GPa, regardless of the indentation size (Fig. 3a). On the other hand, the elastic modulus obtained from the cross-section samples decreases with the increasing penetration depth, showing an ISE (Fig. 4a). The elastic modulus decreases from about 130 GPa at 200 nm indentation depth to about 60 GPa at 1400 nm indentation depth for samples FS1 and FS2 and to about 90 GPa for samples FS3 and FS4. In addition, the ISE of the elastic modulus measured is more obvious in FS1 and FS2 than in FS3 and FS4. In other words, the samples with smaller thickness (FS1 and FS2) show more pronounced ISE of elastic modulus. Figure 5 shows the ISE of both the elastic modulus and hardness measured on the cross section samples with austenitic substrate (AS1-4). Similarly it can be seen that the ISE becomes more pronounced when the film thickness decreases.

The ISE of hardness has been widely reported and well documented from the nanoindentation tests of a large variety of crystalline materials [11, 12] which has been explained using the strain gradient mechanism (see [13, 14] for reference). Although its grain size is generally smaller compared with those testing materials studied in the literature on the ISE, the copper films used in the present study shall be subjected to a mainly dislocation mediated plastic deformation during nanoindentation tests, as the crossover to the grain boundary dominated plastic deformation would occur but at much smaller grain size regime [15].

Different from the hardness, elastic modulus reflects the inter-atomic bonding exists in the testing materials and has little dependence on the microstructure. This is true when measured on the plan view samples despite the different grain sizes. However, the elastic modulus shows an ISE when measured on cross section samples. This may be caused by the fact that the indentation induced plastic deformation has been limited in the film. Due to the specific geometry of the Berkovich indenter tip, an indentation depth of 1000 nm would easily generate an imprint with horizontal length 5~6 μm. The volume of materials involved in the elasto-plastic indentation deformation will cover a significant portion of the thickness or even beyond, if the film thickness is small. Hence the volume of the testing material plays a more significant role when tested on the cross section sample in comparison to that on the plan view sample. This is also in agreement with the observation that the ISE of elastic modulus is more pronounced on samples with smaller thickness (see Fig. 3a and Fig. 4a). Therefore, the ISE of the elastic modulus observed might be regarded as a sample size effect.

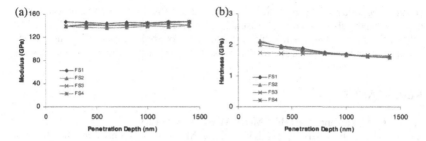

Figure 3. The elastic modulus (a) and hardness (b) as function of penetration depth of the copper film FS1-4 taken from the plan view direction.

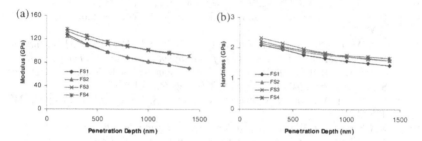

Figure 4. The elastic modulus (a) and hardness (b) as function of penetration depth of the copper film FS1-4 taken from the cross-sectional direction.

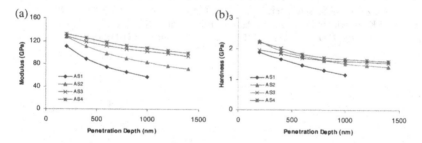

Figure 5. The elastic modulus (a) and hardness (b) as function of penetration depth of the copper film AS1-4 taken from the cross-sectional direction.

CONCLUSIONS

The mechanical properties of electrodeposited copper films of varying thickness and grain size were characterised using nanoindentation technique. The hardness measured shows an ISE. The elastic modulus measured on plan view samples is fairly constant regardless of the grain size and sample thickness. However, an ISE has been observed on the elastic modulus measured on the cross section sample, regardless of the substrate. The film with smaller thickness shows a more pronounced ISE of the elastic modulus. This ISE of the elastic modulus may be regarded as a sample size effect.

REFERENCES

1. A. Al-Rub and K. Rashid, Mech. Mater. **39**, 787 (2007).
2. C.H. Seah, S. Mridha, and L.H. Chan, J. Mater. Proc. Technol. **89-90,** 432 (1999).
3. 'Nanostructured Coatings', edited by. A. Cavaleiro and J.T.M.D. Hosson, Springer, New York (2006).
4. Y. Choi, K.J. Van Vliet, J. Li, and S. Suresh, J. Appl. Phys. **94**,6050 (2003).
5. S. Tao and D.Y. Li, Nanotechnology. **17,** 65 (2006).
6. T.H. Fang and W.J. Chang, Microelect. Eng. **65,** 231 (2003).
7. D. Beegan, S. Chowdhury, and M.T. Laugier, Surf. Coat. Technol. **176,** 124 (2003).
8. A.A. Volinsky, J. Vella, I.S. Adhihetty, V. Sarihan, L. Mercado, B.H. Yeung, and W.W. Gerberich. in *Fundamentals of Nanoindentation and Nanotribology II,* edited by S. P. Baker, R. F. Cook, S. G. Corcoran, and N. R. Moody (Mater. Res. Soc. Symp. Proc. **649**, Boston, MA, 2000) pp. Q.5.3.1-5.3.6.
9. C.Y. Chan, W.J. Zhang, S. Matsumoto, I. Bello, S.T. Lee, J. Crystal Growth, **247,** 438 (2003).
10. X. Chen and J.J. Vlassak. in *Fundamentals of Nanoindentation and Nanotribology II,* edited by S. P. Baker, R. F. Cook, S. G. Corcoran, and N. R. Moody (Mater. Res. Soc. Symp. Proc. **649**, Boston, MA, 2000) pp. Q.1.3.1-1.3.6.
11. G. Feng and W.D. Nix, Scripta Mater. **51,** 599 (2004).
12. I. Manika and J. Maniks, Acta Mater. **54,** 2049 (2006).
13. M. Zhao, W.S. Slaughter, M. Li, S.X. Mao, Acta Mater. **51,** 4461 (2003).
14. N.K. Mukhopadhyay and P. Paufler, Inter. Mater. Rev. **51,** 209 (2006).
15. J. Schiøtz, F.D. Di Tolla and K.W. Jacobsen, Nature **391** (1998).

Mater. Res. Soc. Symp. Proc. Vol. 1049 © 2008 Materials Research Society   1049-AA03-06

Mechanical Properties of 3C-SiC Films for MEMS Applications

Jayadeep Deva Reddy[1], Alex A. Volinsky[1], Christopher L. Frewin[2], Chris Locke[2], and Stephen E. Saddow[2]

[1]Department of Mechanical Engineering, University of South Florida, 4202 E. Fowler Ave. ENB118, Tampa, FL, 33620

[2]Department of Electrical Engineering, University of South Florida, 4202 E. Fowler Ave. ENB118, Tampa, FL, 33620

ABSTRACT

There is a technological need for hard thin films with high elastic modulus and fracture toughness. Silicon carbide (SiC) fulfills such requirements for a variety of applications at high temperatures and for high-wear MEMS. A detailed study of the mechanical properties of single crystal and polycrystalline 3C-SiC films grown on Si substrates was performed by means of nanoindentation using a Berkovich diamond tip. The thickness of both the single and polycrystalline SiC films was around 1-2 μm. Under indentation loads below 500 μN both films exhibit Hertzian elastic contact without plastic deformation. The polycrystalline SiC films have an elastic modulus of 457 $\pm$ 50 GPa and hardness of 33.5 $\pm$ 3.3 GPa, while the single crystalline SiC films elastic modulus and hardness were measured to be 433 $\pm$ 50 GPa and 31.2 $\pm$ 3.7 GPa, respectively. These results indicate that polycrystalline SiC thin films are more attractive for MEMS applications when compared with the single crystal 3C-SiC, which is promising since growing single crystal 3C-SiC films is more challenging.

INTRODUCTION

The development of SiC as a microelectronic material for over 2 decades has resulted in enormous prospects for its use in MEMS applications [1, 2]. Mechanical properties of thin films play a pivotal role in determining the lifetime of MEMS devices. We studied cubic SiC (3C-SiC) films grown on Si substrates, which are chemically inert, can withstand high temperatures, and have a high resistance to oxidation. Silicon carbide also has excellent electronic and thermal properties, including large reverse breakdown voltage, high electron mobility, high saturated electron drift velocity and excellent thermal conductivity relative to Si [3], making SiC attractive for MEMS applications under hostile conditions. Present trends indicate an increasing interest in the cubic polycrystalline form of SiC, namely poly-3C-SiC, as a MEMS material since it can be deposited on various substrates and micromachined in a similar fashion to Si [4].

Silicon carbide exists in more than 200 polytypes, but only a few are useful in fabricating semiconductor devices, of which 3C-SiC has been widely developed for semiconductors. Since silicon carbide can exist in both polycrystalline and single crystal forms, it is important to compare which film is best suited for MEMS applications by measuring their relative mechanical properties. There are various methods used in determining the mechanical properties of thin films, including the bulge test, micro-beam bending, the micro tensile test [2], nanoindentation, etc. In this paper nanoindentation was used to characterize the mechanical properties of single crystal and polycrystalline 3C-SiC films grown on Si substrates in order to shed light on which is better suited for MEMS applications, which require good mechanical properties such as hardness, elastic modulus and fracture toughness.

EXPERIMENT

Single crystal (3C-SiC) and polycrystalline (poly-3C-SiC) SiC samples were grown on Si (100) substrates. The samples were grown heteroepitaxially via chemical vapor deposition (CVD) in a hot-wall reactor [3]. The thickness of the samples was around 1-2 μm. The crystallographic orientations of the samples were not taken into consideration for the mechanical properties measurements since one of the films was polycrystalline. The orientation of the single crystal SiC film was (100).

Growth of single crystal 3C-SiC films

3C-SiC single crystal films were grown on a 50 mm diameter Si (100) wafer using a hot-wall CVD process. Details of both the reactor and the growth process can be found elsewhere [5]. The 3C-SiC on Si deposition process was developed using the two step carbonization and growth method. C_3H_8 and SiH_4 were used as the precursor gases to provide carbon and silicon, respectively. Ultra-high purity hydrogen, purified in a palladium diffusion cell, was employed as the carrier gas. Prior to growth, the Si substrates were prepared using the standard RCA cleaning method [6], followed by a 30 second immersion in diluted hydrofluoric acid to remove surface contaminants and the surface oxide. The first stage of the process, known as the carbonization step, involved heating the Si substrate from room temperature to 1140 °C at atmospheric pressure with a gas flow of 6 standard cubic centimeters per minute (sccm) of C_3H_8 and 10 standard liters per minute (slm) of H_2. The temperature was then maintained for two minutes to carbonize the substrate surface. After carbonization, SiH_4 was introduced into the system at 4 sccm and the temperature was increased to the growth temperature of 1375 °C, and the pressure was changed to 100 Torr for approximately 5 minutes. The temperature and gas flow rates were then maintained at these values during the growth process. By this procedure, a 2 μm thick 3C-SiC film was grown [7]. After the growth process was completed, the wafer was cooled to room temperature in an Ar atmosphere [3]. After deposition the film thickness was measured by Fourier transform infrared reflectance spectroscopy (FTIR) and confirmed by cross-section scanning electron microscopy analysis. The crystal orientation of the grown film was determined by X-ray diffraction (XRD) using a Philips X-Pert X-ray diffractometer. XRD data showed that the films were single crystal with an X-ray rocking curve FWHM of approximately 300 arcsec. Figure 1 shows the rocking curve for the single crystal 3C-SiC film.

Growth of polycrystalline 3C-SiC films

Polycrystalline growth follows the same procedure as single crystal growth with the exception of a higher flux of the growth species. The process conditions for the samples studied here were therefore identical to those listed above except that the SiH_4 and C_3H_8 mass flow rates were 300 and 8.5 sccm, respectively. This process resulted in a poly-3C-SiC film with a thickness of approximately 4 μm.

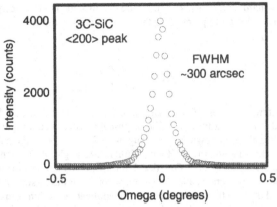

Figure 1. Rocking curve from the (200) plane for the 3C-SiC film grown on (100) Si, confirming that the film is single crystal 3C-SiC.

Nanoindentation procedure

Nanoindentation experiments were performed using a Hysitron Triboindenter. A Berkovich indenter was used to perform all indentation tests. Load controlled indentations were done to determine the elastic modulus and hardness of the films. The indenter tip was calibrated using a quartz standard to determine the tip area function [8]. The radius of the Berkovich tip was approximately 100 nm, which was also confirmed by using the Hertzian theory of elastic contact [9, 10].

The single crystal 3C-SiC sample had an optically smooth (mirror) surface requiring no polishing prior to nanoindentation. However, the polycrystalline sample had a rough morphology which required mechanical polishing. Polycrystalline 3C-SiC films of the same approximate thickness were prepared on Si substrates from the same wafer lot to conduct indentation experiments. The sample was polished on a polishing pad using 1 μm diamond paste to both thin down the thickness of the film to match the single crystal film thickness and to smooth out the film surface roughness; the resulting film thickness was equivalent to the single crystal SiC film (1-2 μm). Samples were then cleaved and glued to the sample holders using Super glue and placed on the indenter stage for measurement. Each indent was performed individually after scanning the sample surface to ensure its local smoothness.

The low load transducer, which can apply a maximum load of 10 mN, was used to determine the elastic modulus (E) and hardness (H) of the 3C-SiC films. The experiment was carried out at loads varying from 0.5 to 10 mN on both the single crystal and polycrystalline SiC samples. To determine the fracture toughness (K), the low load transducer was replaced with a high load transducer. The above mentioned indentation procedure was followed at higher loads ranging from 100 to 1500 mN.

Small cracks on the surface of the thin films were induced when higher loads were applied. These patterns of cracks were used to estimate the fracture toughness of the films. Cracks come in different morphologies depending on the indentation load, tip indenter geometry and the material properties. The most common types of cracks observed are typically radial cracks for brittle and hard materials [11, 12]. Figure 4 shows the load induced radial cracks propagating from the indentation point using the Berkovich tip. Fracture toughness is calculated by using equation (1),

$$K_C = A \sqrt{\frac{E}{H}} \frac{P_{max}}{C^{1.5}} \tag{1},$$

where K_C is the fracture toughness, P_{max} is the maximum load, C is the crack length, E is the elastic modulus and H is the hardness. A is the empirical constant determined from conducting indentation into material with known fracture toughness. It is 0.039 for the cube corner indenter and 0.022 for the Vickers indenter [13]. Having similar face angle, Berkovich tip would have values similar to the Vickers indenter. Equation 1 was originally derived for bulk materials, as it assumes a half-penny shaped crack, and it does not account for the elastic moduli mismatch between the film and the substrate materials. It is not applicable for thin films on hard substrates, where the cannel crack is constrained and does not kink into the substrate. In case of the SiC film on the Si substrate the situation is different, where surface crack in SiC is likely to kink into the Si substrate, thus being half-penny shaped.

Wear tests on the single crystal 3C-SiC film were also performed in a 3 x 3 μm area using the low load transducer at 2 μN normal load and 1 Hz frequency. The objective of this procedure was to determine the wear resistance of the sample by repeated scanning of the poly-3C-SiC surface. After a certain number of wear cycles the scans area was zoomed out to 5 x 5 μm to determine the material wear.

DISCUSSION

Load-controlled indentations were performed to varying maximum load, ranging from 0.5 to 10 mN. The load-displacement curves exhibited by each type of 3C-SiC material are compared in Figure 2 (a) and (b). The hardest material had less penetration depth of the tip into the sample surface and hence the graph shows less displacement for the same amount of load applied.

From Figure 2 it can be inferred that at lower load (<500 μN) both the single and polycrystalline samples made elastic contact with the diamond probe. Low load curves help in determining the radius of the indenter tip used in performing the nanoindentation experiments, by using the Hertz theory of elastic contact [9, 10]. At higher loads varying from 5 to 10 mN plastic deformation was observed in the film. Figure 2 (b) shows the indentation done at a load of 10 mN, from which it can be inferred that the indenter penetrated more into the single crystal 3C-SiC film surface compared to that of the polycrystalline 3C-SiC film.

Figure 3 shows the modulus and hardness values of respective 3C-SiC films. Thin film elastic modulus (E_{sample}) obtained by nanoindentation is calculated from:

$$\frac{1}{E_r} = \frac{1 - \nu_{sample}^2}{E_{sample}} + \frac{1 - \nu_{tip}^2}{E_{tip}} \tag{2},$$

where $E_{tip} = 1140$ GPa and $v_{tip} = 0.2$ are the elastic modulus and Poisson's ratio for the diamond tip, respectively.

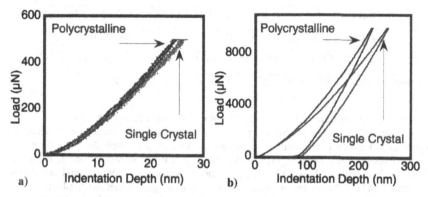

Figure 2. Load-displacement curves at the maximum load of a) 1 mN and b) 10 mN for polycrystalline and single crystal 3C-SiC films grown on (100) Si.

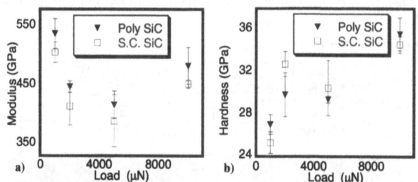

Figure 3. a) Elastic modulus and b) Hardness of single crystal and poly crystalline 3C-SiC films grown on (100) Si. Corresponding maximum indentation depths are 60, 80, 160 and 250 nm, respectively.

Table I reports the average values of the measured hardness and the elastic modulus of single crystal 3C-SiC, polycrystalline 3C-SiC, a 15R-SiC Lely platelet and Si (100) substrate. At least five indents were performed at each maximum load. 15R-SiC was used as a comparison since this material is known to have a minimum amount of dislocations. There is a relatively large absolute variation in measured elastic modulus (±50 GPa), not typically observed in indentation of softer materials. One has to consider the relative variation, which is typically on the order of 10%. The reported mechanical properties are for a maximum indentation depth range between 60 and 250 nm, where quartz calibration is reasonable.

Table I. Measured SiC mechanical properties.

Material	Hardness (GPa)	Elastic Modulus (GPa)
Silicon (100)	12.46 ± 0.78	172.13 ± 7.76
Lely platelet 15R-SiC	42.76 ± 1.19	442 ± 16.34
Single crystal 3C-SiC on (100) Si	31.198 ± 3.7	433 ± 50
Polycrystalline 3C-SiC on (100) Si	33.54 ± 3.3	457 ± 50

Figure 4 shows the microscopic images of cracks induced at higher loads in polycrystalline and single crystal 3C-SiC, respectively. The crack lengths were used to calculate the film fracture toughness using equation (1). Radial cracks were generated along the sharp corners of the Berkovich tip. Table II gives fracture toughness values of respective SiC films on the Si substrate along with the bulk SiC fracture toughness measured by conventional techniques and nanoindentation [14]. Normally nanoindentation fracture toughness method employing equation 1 gives 40% variability compared with other techniques. The cause for the low fracture toughness in our case compared to the bulk values is due to the tip penetrating into the Si substrate and substrate fracture skewing the results to a lower substrate fracture toughness value. Note that for the single crystal 3C-SiC film radial cracks are aligned along the Si substrate crystallographic orientation (Figure 4b) opposite to typical radial cracks in the poly-3C-SiC in Figure 4a.

Table II. SiC fracture toughness.

Material	Fracture Toughness (MPa·m$^{1/2}$)
Single crystal 3C-SiC film on Si	1.59 ± 0.21
Polycrystalline 3C-SiC film on Si	1.54 ± 0.28
Bulk SiC	4.6
Bulk SiC (Eq. 1) [14]	3.1 ± 0.9

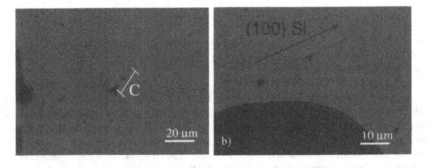

Figure 4. Micrographs of a) Poly-3C-SiC film on (100) Si substrate indent at 1500 mN and b) 3C-SiC film on (100) Si substrate indent at 1000 mN.

Since it was determined that poly-3C-SiC film has higher hardness and elastic modulus, thus is better suited for MEMS applications, it was also tested for wear resistance. Figure 5 shows the topography images and corresponding line profiles of the polycrystalline 3C-SiC surface before and after tip-induced wear test performed for 1045 scans. During scanning there was very little or negligible wear, as only 1-2 nm of material depth was removed. This result confirms the high wear resistance of poly-3C-SiC films necessary for MEMS applications.

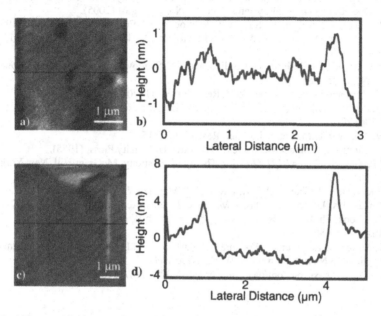

Figure 5. a) Topography image of the original poly-3C-SiC surface and b) Line scan showing initial surface roughness; c) Wear square area after 1045 scans and d) Corresponding line profile showing about 2 nm of wear depth.

CONCLUSIONS

3C-SiC thin films grown on (100) Si for MEMS applications were observed to be very hard and have high elastic modulus compared to the silicon wafers. The nanoindentation results show that polycrystalline SiC films have better mechanical properties compared to single crystal 3C-SiC films and should be suitable for MEMS applications in harsh environments.

ACKNOWLEDGEMENTS

The 3C-SiC growth in S.E. Saddow's laboratory was supported by the Army Research Laboratory under Grant No. DAAD19-R-0017 (B. Geil) and the Office of Naval Research under

Grant No. W911NF-05-2-0028 (C. E. C. Wood). Alex Volinsky would like to acknowledge support from NSF under CMMI contracts 0631526, 0600266 and 0600231.

REFERENCES

1. M. Ohring, The Material Science of Thin Films, Academic Press, San Diego, (1992).
2. K.M. Jackson, J. Dunning, C.A. Zorman, M. Meheregany, W.N. Sharpe, Journal of Microelectromechanical Systems, Vol. 14, No. 4, August (2005).
3. R.F. Davis, G. Kelner, M. Shur, J.W. Palmour and J. Edmond, Proc. IEEE 79 (1991) 677.
4. C.A. Zorman, S. Roy, C.H.Wu, A.J. Flieschman, M. Mehregany, J. Mater. Res.13 (1998) 406.
5. R. L. Myers, Y. Shishkin, O. Kordina, and S.E. Saddow, *Journal of Crystal Growth* vol. 285/4, (2005) 483-486.
6. W. Kern and D. A. Puotinen, RCA Rev. Vol. 31, (1970) 187-206.
7. M. Reyes, Y. Shishkin, S. Harvey, S.E. Saddow, Mat. Res. Soc. Symp. Proc. Vol. 911, (2006) 79.
8. W.C. Oliver, G.M. Pharr, J. Mater. Res. 7 (1992) 1562.
9. K.L. Johnson, Contact Mechanics, Cambridge University Press, (1985).
10. S.P. Timoshenko and J.N. Goodier, Theory of Elasticity, McGraw-Hill, New York, (1970).
11. G.R. Anstis, P. Chantikul, B.R. Lawn, D.B. Marshall. J. Am. Ceram. Soc (1981). 64:533.
12. D. Casellas, J. Caro, S. Molas, J. M. Prado, I. Valls, Acta Mater. 55 (2007) 4277-4286.
13. R.F. Cook, Ph. D. Thesis, University of New South Wales, Sydney, New South Wales, Australia, 1986.
14. D. Yang and T. Anderson, Fracture Toughness of SiC Determined by Nanoindentation, Application Note, Hysitron Inc. World Wide Web: http://www.hysitron.com/page_attachments/0000/0367/111500-001.pdf

Mater. Res. Soc. Symp. Proc. Vol. 1049 © 2008 Materials Research Society   1049-AA03-07

Strength Measurement in Brittle Thin Films

Oscar Borrero-Lopez[1,2], Mark Hoffman[1], Avi Bendavid[3], and Phil J Martin[3]
[1]University of New South Wales, Sydney, NSW 2025, Australia
[2]Universidad de Extremadura, Badajoz, 06071, Spain
[3]Materials Science and Engineering, CSIRO, Lindfield, NSW 2070, Australia

ABSTRACT

In this work we have investigated the strength variability of brittle thin films (thickness $\leq$ 1 μm) utilising a simple test methodology. Nanoindentation of as-deposited tetrahedral amorphous carbon (ta-C) and Ti-Si-N nanocomposite films on silicon substrates followed by cross-sectional examination of the damage with a Focused Ion Beam (FIB) Miller allows the occurrence of cracking to be assessed in comparison with discontinuities (pop-ins) in the load-displacement curves. Strength is determined from the critical loads at which cracking occurs using the theory of plates on a compliant foundation. This is of great relevance, since the fracture strength of thin films ultimately controls their reliable use in a broad range of functional applications.

INTRODUCTION

Brittle thin films (thickness $\leq$ 1μm) are one of the most commonly used means of overcoming materials-based limitations to attain increased functionality in components. Examples of their use include tribological films for wear parts, magnetic drives, biomedical applications, or MEMS devices. The mechanical properties of thin films that have been most widely studied are elastic modulus and hardness, using nanoindentation probe techniques. However, the fracture strength of the films, which ultimately limits their structural integrity and reliable application, has received much less attention.

While in bulk specimens the procedure for strength measurement is well-established —*i.e.*, three- or four-point bending, biaxial flexure tests, and so forth—, direct application of these experiments at the sub-micron scale is not straightforward. Current methodologies for strength measurement in brittle thin films and small systems relevant to MEMS consist in nanomechanics testing of micromachined free-standing films [1-6]. However, these methods lack simplicity and present limitations. Namely (i) only one single datum per specimen can be obtained, (ii) micromachining techniques may induce defects in the resulting specimens, which in turn can influence the measured strength, and (iii) the use of free-standing films is limited to nearly stress-free systems, as residual stresses hinder machinability of microbeams.

Based on the above, the aim of this work is to extrapolate the use of a simple procedure developed by Lawn *et al.* for thick coatings to the realm of sub-micron films [7, 8]. Using this methodology, films can be tested as-deposited in the substrate avoiding micromachining. Furthermore, it allows films with high residual stresses to be tested, and several tests to be performed within the same specimen.

EXPERIMENTAL

Materials

Filtered-arc deposited tetrahedral amorphous carbon (ta-C) and arc-magnetron deposited Ti-Si-N nanocomposite films (thickness 500 nm and Young's modulus 300 GPa) on silicon substrates (Young's modulus 170 GPa) are considered to prove the applicability of the proposed methodology. Diamond-like carbon (DLC; thickness 1 μm and Young's modulus 160 GPa, *i.e.* lower fraction of sp^3 bonds compared to ta-C) is employed in both systems as a buffer layer, as shown in Fig. 1. Note that the buffer layer is only an artifact to be able to measure the strength of the thin films (ta-C and Ti-Si-N nanocomposites). FEA simulations show that the use of the buffer layer ensures maximum tensile stresses in the films [9].

Nanoindentation

As-deposited systems were subjected to nanoindentation in an Ultra-Micro Indentation System UMIS 2000 (CSIRO, Sydney, Australia). A spherical-tipped conical indenter of radius 5 μm was used to apply maximum loads of up to 500 mN.

Microscopy

The Focused Ion Beam Mill (200xP, FEI Company, Hillsboro, OR) was used to prepare and image cross-sections of indentation sites in order to investigate the fracture in the different systems. High beam currents (1000 pA and 350 pA) were used to mill and clean the cross-sections. Subsequently, the stage was tilted and the cross-sections were viewed at 45° angle using a low beam current (10 pA).

RESULTS

According to the works by Lawn *et al.* and Kim *et al.* [7, 8, 10], the fracture strength of a brittle coating (σ_c) on a compliant substrate can be determined from the critical load at which cracking in the coating initiates (P_{Crit}) upon progressive loading of the system with a spherical indenter (see Fig. 1), using the theory of plates on a compliant foundation[1] (Eq. (1)). In Eq. (1), d^*_C and E^* are the effective thickness and Young's modulus of the coating, respectively; E_S is the Young's modulus of the substrate and C and B^* are constants. In thick coatings, critical loads are directly determined by monitoring the system from below the substrate with a CCD camera. However, the small size in thin films precludes direct crack observation, and crack initiation has to be assessed indirectly from discontinuities (pop-ins) in the nanoindentation load-displacement curves [9].

$$\sigma_C = \frac{P_{Crit}}{B^* d_C^{*2}} \log\left(\frac{CE^*}{E_s}\right)$$

Eq. (1)

---

[1] It is assumed that the stress field remains linear elastic up to the point of fracture.

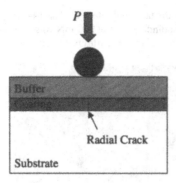

Figure 1. Schematic illustrating the cross-section of a trilayer system with a thin coating (ta-C or Ti-Si-N) and a buffer layer (DLC) loaded with a spherical indenter.

Figure 2 shows characteristic nanoindentation load-displacement curves corresponding to the ta-C system with and without buffer layer, and the bare Si substrate. Different pop-ins can be observed in the three curves, typically associated to fracture events. Based on a FE analysis [9], the first pop-in in the trilayer system (with the buffer layer) should correspond to the first fracture event in the ta-C film. Note that this pop-in can not be detected when the buffer layer is not employed. Furthermore, no pop-in at similar contact pressure can be noticed in the bare silicon.

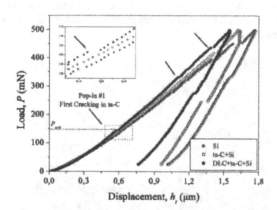

Figure 2. Nanoindentation load-displacement curves corresponding to the ta-C system. Arrows point bursts in the trilayer system curve (pop-ins), indicative of fracture events. The inset shows a magnified detail of the first pop-in in the trilayer system.

To confirm that fracture of the ta-C took place at P_{crit} (144 mN), cross-sectional FIB examination of the system after nanoindentation at 200 mN was carried out, and is shown in Fig. 3, where fracture in the lower surface of the ta-C film is evident.

Figure 3. Cross-sectional FIB image of the damage in the ta-C trilayer system after nanoindentation at a maximum load of 200 mN. Cracking in the ta-C film can be observed in the micrograph.

At loads higher than P_{crit}, more pop-ins appear in the load-displacement curve, indicative of further fracture processes, as shown in Fig. 4 after indentation at 400 mN. The microscopy analysis confirms that the load at which the first pop-in in the load-displacement curve is detected corresponds to the critical load at which fracture of the ta-C film takes place. Inserting the value of P_{crit} in Eq. (1), allows the fracture strength of the film to be easily estimated. By simply performing 30 nanoindentation experiments at different locations of the same sample, Weibull plots can be readily obtained.

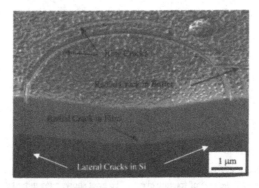

Figure 4. Cross-sectional FIB image of the damage in the ta-C trilayer system after nanoindentation at a maximum load of 400 mN.

The same methodology can be easily applied to other thin film systems. In particular, in this work we also consider a Ti-Si-N nanocomposite film. Figure 5 shows the Weibull plots for both the ta-C and the Ti-Si-N films. A characteristic strength of 1.7 GPa and a Weibull modulus of 10 were obtained for the ta-C film, which compares well with results obtained by Espinosa *et al.* from nanoindentation of fixed-end microbeams [2]. For the Ti-Si-N nanocomposite, the characteristic strength obtained is 4.3 GPa, about 2.5 times higher than in the ta-C case. In addition, the nanocomposite shows higher reliability (Weibull modulus of 18).

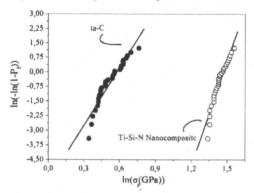

Figure 5. Weibull plots of the ta-C and Ti-Si-N films (thickness 500 nm). P_f is the probability of failure and σ_F is the fracture strength. The points correspond to experimental data and the solid lines are the best linear fits to the experimental data

CONCLUSIONS

The fracture strength of sub-micron brittle films can be obtained from discontinuities (pop-ins) in the nanoindentation load-displacement curves of the as-deposited films, using the theory of plates on a soft foundation. This methodology presents multiple advantages over pre-existing micromechanical testing methods of free-standing films. Namely, it involves simpler instrumentation, allows several tests per sample to be performed, and avoids specimen micromachining, which permits films with high residual stresses to be tested.

A characteristic strength of 1.7 GPa and a Weibull modulus of 10 were obtained for a ta-C film of 500 nm thickness, which compares well with previous results obtained with free-standing films.

The characteristic strength and Weibull modulus of a Ti-Si-N nanocomposite film were measured for the first time. The values obtained are 4.3 GPa and 18, respectively.

REFERENCES

1. I. Chasiotis and W. G. Knauss, Exp. Mech. 42 (2002) 51.
2. H. D. Espinosa, B. Peng, N. Moldovan, T. A. Friedmann, X. Xiao, D. C. Mancini, O. Auciello, J. Carlisle, C. A. Zorman and M. Merhegany, Appl. Phys. Lett. 89 (2006) 073111.
3. T. Namazu, Y. Isono and T. Tanaka, J. Microelectromech. S. 9 (2000) 450.
4. W. N. Sharpe, Jr., B. Yuan and R. L. Edwards, J. Microelectromech. S. 6 (1997) 193.
5. S. Sundararajan and B. Bhushan, Sensor Actuat. A-Phys. 101 (2002) 338.
6. T. Tsuchiya, O. Tabata, J. Sakata and Y. Taga, J. Microelectromech. S. 7 (1998) 106.
7. H. Chai, B. Lawn and S. Wuttiphan, J. Mater. Res. 14 (1999) 3805.
8. Y. W. Rhee, H. W. Kim, Y. Deng and B. R. Lawn, J. Am. Ceram. Soc. 84 (2001) 1066.
9. O. Borrero-López, M. Hoffman, A. Bendavid and P. J. Martin, Acta Mater. In Press (2008).
10. J. H. Kim, H. K. Lee and D. K. Kim, Philos. Mag. 86 (2006) 5383.

Nanotribology and Friction

Mater. Res. Soc. Symp. Proc. Vol. 1049 © 2008 Materials Research Society   1049-AA04-04

Measurement of Ultrathin Film Mechanical Properties by Integrated Nano-scratch/indentation Approach

Ashraf Bastawros[1,2], Wei Che[3], and Abhijit Chandra[1,2]

[1]Aerospace Engineering, Iowa State University, Ames, IA, 50011
[2]Mechanical Engineering, Iowa State University, Ames, IA, 50011
[3]Saint Gobain, Inc., Boston, MA, 01606

ABSTRACT

The thickness and property measurements of thin films on substrates are crucial for wide range of applications. Classical techniques have relied on various physical properties to identify film thickness, independent of its mechanical properties. Here, a new experimental technique is devised to evaluate the film thickness, its flow stress and its stiffness. The technique utilizes a combination of nano-scratch and dynamic stiffness measurements carried out by a nano-indenter. The technique relies on measuring the depth variation of normal and tangential forces, and indentation modulus. These combined measurements are calibrated through a simple statically admissible stress model to yield the unknown quantities. The measurements are ascertained by XPS film thickness measurements, and reasonably agree with the finite element predictions. The technique is applied to study the formed oxide nano-layer during copper chemical mechanical planarization.

INTRODUCTION

Ultrathin films are widely used in many applications such as dielectric films, copper liners and capping materials for microelectronics; native oxides for high temperature alloys, or evolving chemical product layer in chemical mechanical planarization (CMP) [1]. The physical size limitation of thin/ultra-thin films prohibits direct measurements of the film mechanical properties, except in some special circumstances (e.g. blanket wafer curvature to measure film stresses [2], or blister test to probe monolithic film properties [3]). Nano-indentation, on the other hand, provides a convoluted data, which depends on several sets of the film mechanical properties as well as its thickness [4, 5]. We are proposing a novel approach to measure the mechanical properties of ultra-thin (relatively soft) on a substrate (relatively hard), through several sets of measurements, combining nano-scratch and continuous stiffness measurements through the thickness of the film and up to an indentation depth of 8-10 and scratch depth of 2-3 times the film thickness. These measurements are analyzed by the simple limit analysis model and compared against a single phenomenological correlation, calibrated by finite element analysis.

PROPOSED METHODOLOGY

The proposed methodology utilizes the imposed deformation field by a nano-scratch to determine the film thickness and its mechanical properties. A sketch of a nano-scratch is depicted in Fig. 1, which ploughs the entire thickness of the film, t, and continue into the substrate. In the current simplified analysis, full contact is assumed between the indenter and film/substrate combination across the entire depth of indentation. Also, pile-up or sink-in is ignored for the soft film on hard substrate system, examined here. Using lower bound limit analysis, a statically admissible stress field, in equilibrium with the applied load can be assumed. For such a field, the stresses normal to the contact interface are proportional to the yield stress of both the film and substrate, respectively. It should be noted that the constrained deformation at the interface may change the stress triaxiality and the pressure build up underneath the indenter. However, such effect can be safely ignored, since the limited thickness of film may help in relaxing the added constraints along the out of plane direction via the free surface.

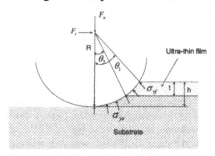

Figure 1. Sketch of the stress distribution below a traveling indenter.

For frictionless indenter and shallow indentation, the normal, F_n and tangential F_t forces, which are in equilibrium with the assumed stress state can be written in the form:

$$F_n \approx \pi \xi \sigma_{ys} \left(R(h-t) + \frac{\sigma_{yf}}{\sigma_{ys}} Rt \right) \tag{1}$$

$$F_t \approx \frac{4\sqrt{2R}}{3} \xi \sigma_{ys} \left((h-t)^{3/2} + \frac{\sigma_{yf}}{\sigma_{ys}} \left(h^{3/2} - (h-t)^{3/2} \right) \right) \ . \tag{2}$$

Here h is the scratch depth, σ_{yf} and σ_{ys} are the film and substrate yield strength, respectively. ξ is a proportionality constant that correlate the normal stress below the indenter to the film/substrate flow stress, for an assumed statically admissible stress state. Utilizing the ratio F_n/F_t would eliminate the need to know exact values for both ξ and σ_{ys}. For the range of measurements of F_n/F_t ratio for each h, an iterative approach can be utilized to yield the best fit for the film thickness, t, and the yield strength ratio, σ_{yf}/σ_{ys}. The remaining quantity is the modulus ratio, E_{yf}/E_{ys}. Here, it will be estimated from a combination of nano-dynamic stiffness measurements (DSM) to estimate the variation of the combined indentation modulus with the indentation depth, then compare the results with those from finite element simulation (FEM) to estimate the ratio of E_{yf}/E_{ys}.

EXPERIMENT

Electroplated copper film is deposited onto oxygen free, 99.99% high purity copper (101-alloy series) discs. An acid bath of $CuSO_4+5H_2O$ (0.25 mol/L) and H_2SO_4 (1.8 mol/L) is used [6]. A deposition time for 1 hr and 0.5 V is employed to achieve a 50 μm thick Cu layer. The specimen is annealed at room temperature for 24 hrs to achieve 1-2 μm grain size [7]. Finally the specimen is gently polished with 1 and 0.05 μm alumina particles respectively to achieve 3-4 nm roughness. The surface is then exposed to 0.6wt% NH_4OH for 0, 90, 300 and 600 s to generate soft chemical product layer with varying thickness, and then is air dried. Immediately after exposure, four sets of nano-scratches, 5 μm long, are carried out on the etched surface at loads of 10, 20, 30 and 40 μN.

All nano-scratches and nano-DSM are carried out immediately on the chemically exposed surface, utilizing a Hysitron Tribo-indenter with a Berkovich tip having radius of about 400 nm. A set of four nano-scratches, 5 μm each are carried out at loads of 10, 20, 30 and 40 μN. For the nano-DSM, a ramping quasi-static load is utilized for a ramp of 10 to 100 μN. Ten steps are used for the dynamic loading with amplitude of 50% current quasi-static load and frequency of 10 Hz.

RESULTS AND DISCUSSION

The nano-scratch results are summarized in Fig. 2 for the different exposure time. The relative change of the scratch depth, relative to the bare substrate is depicted on Fig. 3. At lower applied forces (10 μN), the relative change is much higher, ranging from 40% at 90 s to 80% at 600 s, which is an indication that most of the scratch is within the product film. At higher applied forces (40 μN), the relative change was about 10% at 90s and 30% at 600 s. Thus the experimental measurements are true manifestation of the film thickening with increasing the exposure time to the chemically active slurry.

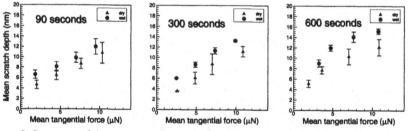

Figure 2. Summary of the nano-scratch results for different exposure time.

Based on the experimental data and simplified nano-scratch model, the experimentally derived film thickness and yield strength are summarized in Table 1. The measurements are quite sensitive to very small variation in film thickness. The average yield strength of the film has a narrow range of 0.45 to 0.52 for the three data sets. It should be noted that these measurements of the copper oxide flow stress are in the hydrated state. Thus the oxide layer has a lower flow stress than the copper substrate.

These film thickness measurements are verified by X-ray photo-electron spectroscopy (XPS). Following the method by Chawla et al [8], angle resolved XPS is used to model the thickness, structure and composition of the film. The details of the analysis can be found in [9]. The XPS results are summarized in Table 2. The range of the estimated film thickness is very similar for the two methods, ascertaining the validity and accuracy of the current proposed method.

Table 1: Experimentally derived film thickness and strength

Exposure time	90s	300s	600s
Film thickness, t (nm)	5.0±0.2	5.7± 0.6	7.3±0.8
σ_{yf}/σ_{ys}	0.52	0.48	0.45

Table 2: XPS film thickness estimates

Exposure time	90s	300s	600s
1st layer (nm)	0.8	0.6	0.5
2st layer (nm)	4.0	5.2	6.6
Film thickness, t (nm)	4.8	5.8	7.1

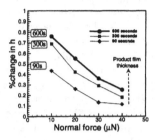

Figure 3. Relative change of scratch depth.

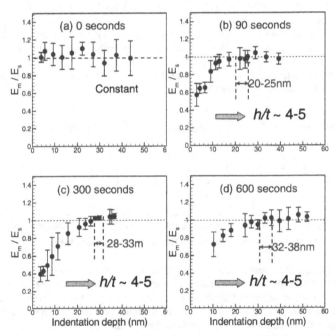

Figure 4. Variation of the measured effective layer modulus by nano-DSM with the indentation depth for different exposure time.

Summary of the nano-DSM effective stiffness measurements, normalized by the substrate modulus, as a function of the indentation depth is summarized on Fig. 4. The measured modulus reaches that of the substrate when the indentation depth, h is about 4-5 times the film thickness. Similar trend has been reported for soft film on hard substrate [10], and indentation depth of about two times the film thickness. The rate of the effective modulus increase until reaching that of the substrate is a strong function of the film to substrate modulus ratio, E_{yf}/E_{ys}. Such dependence can be analyzed numerically via finite element. Figure 5 shows the results of axisymmetric FEM calculation for different E_{yf}/E_{ys}. Details of the calculation can be found in [9]. In this analysis, the indenter is assumed rigid and frictionless. The substrate and film materials are assumed to follow elastoplastic, strain hardening with hardening exponent of 0.2. Since Poisson's ratio was noted in earlier work [11] to have a minor role on the static indentation field, it was kept the same at 0.33 for both the film and substrate. A ratio of σ_{yf}/σ_{ys} =0.45 is utilized. Different cases of E_{yf}/E_{ys} =0.1-1 are examined. For each case, incremental loading/unloading are carried out. The slope of the unloading curve, S, is used to estimate an effective contact stiffness using Oliver-Pharr method [12].

For E_{yf}/E_{ys} =0.5, the substrate modulus is reached at h/t=2. While for E_{yf}/E_{ys} =0.1 the substrate modulus is not reached until h/t>8. Again, this is in agreement with reported work [10] wherein the effective modulus was continuously varying for h/t=2 range and E_{yf}/E_{ys} =1/6. By comparing the measured trend in Fig. 4, with the parametric curves of Fig. 5, the E_{yf}/E_{ys} ratio is found to be about 0.2.

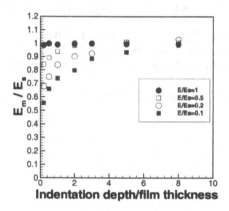

Figure 5. Summary of the FEM parametric study, showing the variation of the film/substrate effective modulus with the indentation depth (normalized by the film thickness).

CONCLUSIONS

The proposed experimental protocol showed promising results to evaluate the film thickness, flow stress ratio and modulus ratio for relatively soft film on a hard substrate. The results are independent of the assumed equilibrium stress state. Further analysis is needed for the generalization of the method to account for extensive pile-up during the nano-scratch experiment as well as for other combinations of hard films on relatively soft substrates.

ACKNOWLEDGMENTS

This work is supported by US-National Science Foundation NSF through grant No. CMMI-0134111.

REFERENCES

1. Steigerwald J.M., Murarka S.P., Gutmann R.J., Duquette D.J., Materials Chemistry and Physics, v 41, n 3, Aug 1995, p 217.
2. Flinn P.A., Gardner D.S., Nix W.D., IEEE Transactions on Electron Devices, v ED-34, n 3, Mar, 1987, p 689-699.
3. Anderson A.P., Devries K.L., Williams M.L., Int. J. Fract. 10 (1974), p. 565.
4. Tsui T.Y., Ross, C.A.; Pharr, G.M., Materials Research Society Symposium - Proceedings, v 473, Materials Reliability in Microelectronics VII, 1997, p 57-62.
5. Tsui, T.Y., Pharr, G.M., Journal of Materials Research, v 14, n 1, Jan, 1999, p 292-301.
6. Weisenberger L.M., DurKin B.J., Copper Plating, Electroplating, pp 167-175, McGraw-Hill, 1978.
7. Gignac L.M., Rodbell K.P., Cabral C. Jr., Andricacos P.C., Rice P.M., Beyers R.B., Locke P.S., Klepeis S.J., Materials Research Society Symposium - Proceedings, v 562, 1999, p 209-214.
8. Chawla S. K., Rickett B. I., Sankarraman N., Payer J. H.,Corrosion Science, Vol. 33, No. 10, 1992, p1617-1631.
9. Che, W., PhD Thesis, Iowa State University, Ames, IA, 2005.
10. Saha R., Nix W.D., *Acta Materialia*, v 50, n 1, Jan 2002, p 23-38.
11. Mesarovic S.D., Fleck N.A., International Journal of Solids and Structures, v 37, n 46, Nov 2000, p 7071-7091.
12. Oliver W.C., Pharr G.M., J Mater Res, 7, 1992, p1564.

Mater. Res. Soc. Symp. Proc. Vol. 1049 © 2008 Materials Research Society   1049-AA04-09

Theory of Lubrication due to Poly-Electrolyte Polymer Brushes

Jeffrey B. Sokoloff

Physics, Northeastern University, 110 Forsyth Street, Boston, MA, 02115

ABSTRACT

It is shown using a method based on the mean field theory of Miklavic Marcelja [1] that it should be possible for osmotic pressure due to the counterions associated with the two polyelectrolyte polymer brush coated surfaces to support a reasonable load (i.e., about 10^5 Pa) with the brushes held sufficiently far apart to prevent entanglement of polymers belonging to the two brushes, thus avoiding what is likely to be the dominant mechanisms for static and dry friction.

INTRODUCTION

Polymer brush coatings on solid surfaces, illustrated in Fig. 1, provide very effective lubrication, in the sense that they are able to support significant load, but have exceedingly low friction coefficients [1]. Human and animal joints are known to exhibit very low friction and wear. The outer surfaces of the cartilage coating these joints have polymeric molecules protruding from them [3–15], which suggests the strong possibility that their very effective lubrication is a result of polymer brush lubrication.

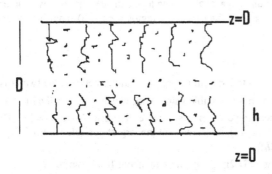

FIG. 1: The geometry of two polyelectrolyte polymer brush coated surfaces with the load pushing the surfaces together supported by osmotic pressure due to counterions in the interface regions separating the tops of the brushes is illustrated schematically. The dots located among the polymer chains and in the interface region between the two brushes represent the counterions. As illustrated here, D denotes the spacing of the surfaces and h denotes the polymer brush height.

Raviv, et. al. [16], have found that polyelectrolyte brushes exhibit remarkably low friction coefficients (10^{-3} or less) compared to the friction coefficients typically found for neutral polymer brushes [2, 17, 18], which already exhibit low friction. In an effort to determine a possible mechanism for this, a solution for the Poisson-Boltzmann equation beyond the Debye-Huckel approximation will be used to determine the concentration of counterions in a region located midway between the two polyelectrolyte brushes. This result is used to show that for polyelectrolyte brushes, osmotic pressure due to the counterions is capable of supporting about 10^5Pa of load, even with the tops of the mean field theoretic monomer profiles of the two brushes about 100A$^\circ$ apart. In fact, it will be argued that this result is valid even for salt concentrations up to 0.1 M, the salt concentration in living matter. As a consequence, there will be little entanglement of the polymers belonging to the two brushes. It was argued in Ref. [19] that such entanglement of polymer brushes leads to static and nonzero slow sliding speed limit kinetic friction, and is likely responsible for wear as well.

THEORY

Ref. [19] shows that the methods of Ref. [1] may be applied to polyelectrolyte brushes to obtain the probability that monomers stick out a distance $z_0 - h$ into the interface region between the two brushes,

$$P_{osm} \propto \exp[-(\frac{z_0 - h}{\xi})^{3/2}] \tag{1}$$

where $\xi / h = (3/\pi)^{3/2} (N^{1/2} a / h)^{4/3}$. Using typical parameters [21]: N=1300, a = 5A$^\circ$ and h = 500A$^\circ$ in the formula for ξ under Eq. (1), we find that $\xi = 100$A$^\circ$. In fact, numerical work on neutral polymer brushes [22, 23] shows that the monomer density is generally quite small in the tails on the monomer density distribution of length ξ, implied by Eq. (1), and there is every reason to assume that the same will be true for charged polymers as well, making it likely that it is not necessary for $z_0 - h$ to be much larger than ξ in order for the monomer density in the tails to be in the dilute regime, in which friction due to blob entanglement does not occur [19].
The possibility will be explored here that a reasonably large load can be supported by counterion osmotic pressure with the plates farther apart than ξ. in the region $0 < z < h$. The electrostatic potential divided by e in the region
$h < z < D - h$ is [24]

$$\psi = \ell_B^{-1} \ln(\cos^2 k_0 (z - D/2)), \tag{2}$$

where $k_0^2 = 2\pi n_0 \ell_B$, where n_0 is the counterion number density midway between the two brushes and ℓ_B is the Bjerrum length. The maximum possible value of k_0, which occurs in the limit of large charge due to both the monomers and counterions inside each brush [24], is $k_0 = [\pi/(D - 2h)]$. From the definition of k_0 below Eq. (2), it follows that the largest possible value of

$$k_0 = [\pi / (D - 2h)] \tag{3}$$

which gives for a value of the parameter D-2h, comparable to about 100A^0, an osmotic pressure $P_{osm} = k_B T n_0 = 0.9 \times 10^5$Pa, and it is inversely proportional to the square of the spacing between the tops of the brushes. In order to verify that n_0 is comparable to the value for it that one obtains from Eq. (3), the Poisson-Boltzmann equation inside a brush, for the case of no excess salt,

$$\frac{d^2 \bar{\varphi}}{dz^2} = -4\pi[n_0 e^{-\bar{\varphi}} - f \varphi(z)] \quad , \tag{4}$$

was solved for $\bar{\varphi} = \ell_B \psi$. In order to make it possible to integrate Eq. (4), we will approximate the monomer density of a brush $\varphi(z)$ by the step function $\varphi(z) = (N/hs^2)\theta(h-z)$, where $\theta(x) = 1$ for $x > 0$ and 0 for $x < 0$ and s is the mean spacing of the polymers on the surfaces. This is a reasonable approximation because we are considering polymer brushes which are compressed because they are supporting a load, and under such circumstances, the parabolic density profile of the uncompressed brush gets flattened into a form that is not too different from the step function form given above [17, 22]. The details of the solution of Eq. (4) are given in Ref. 25, which confirm that P_{osm} is generally of order 10^5Pa.

When there is excess salt present in the solvent, Eq. (4) gets replaced by

$$\frac{d^2\varphi'}{dz^2} = -4\pi\ell_B[n_s(e^{-\bar{\varphi}} - e^{\bar{\varphi}}) - f\varphi(z)] \tag{5}$$

where n_s represents the salt concentration and the second exponential term on the right hand side represents the contribution of ions with the same charge as the brushes to the ionic charge between the plates [5]. For the case of no excess salt, described by Eq. (4), $\bar{\varphi}(z)$ was taken to be zero at z=D/2. In contrast, for the case of excess salt, described by Eq. (5), $\bar{\varphi}(z)$ is zero at $|z| = \infty$. For a low concentration of excess salt, the conditions under which Eq.(5) takes the same form as Eq. (4) will be examined using simple physical arguments, which give results which are identical to those obtained in appendix A from the exact solution to the Poisson-Boltzmann equation of the second article in Ref. [26]. In order to accomplish this, let us write $\bar{\varphi}(z)$ as $\bar{\varphi}(z) = \varphi_0 + \varphi'(z)$ where $\varphi_0 = \bar{\varphi}(D/2)$ and $\varphi'(z) = 0$ is zero at z=D/2. Then, we can make Eq. (4) look like Eq. (6), if we use the fact that $n_0 = n_s e^{-\varphi_0}$. Then we can write Eq. (5) as

$$\frac{d^2\varphi'}{dz^2} = -4\pi\ell_B[n_0(e^{-\varphi'} - \frac{n_s^2}{n_0}e^{\varphi'}) - f\varphi(z)] \tag{6}$$

When $n_s/n_0 = e^{\varphi_0} \ll 1$, Eq. (4) is definitely a good approximation to the problem, and we are justified in treating the system as one without excess salt. To determine the conditions under which Eq. (4) is a good approximation, we solve Eq. (4), in order to determine n_0 as described earlier in this section, and determine φ_0 from $n_0 = n_s e^{-\varphi_0}$, and use φ_0 and the solution of Eq. (4) for $\bar{\varphi}(z)$, which we identify with $\varphi'(z)$ (which is the approximate solution to Eq. (6)) to determine the conditions under which we may neglect $e^{\bar{\varphi}(z)}$ compared to $e^{-\bar{\varphi}(z)}$. (Remember that $\bar{\varphi}(z)$ is negative.) Since it is easily seen from Eq. (2) that $e^{-\varphi'(z)}$ is significantly greater than 1 over much of the range of z from 0 to h for high density polymer brushes, all that is required in order to neglect $e^{\bar{\varphi}(z)}$ in Eq. (4) is that φ_0 be of order unity, which is already satisfied for the 0.1 M salt concentration typical of living matter. It is demonstrated in the appendix of Ref. 25 that these results follow directly from the exact solution of the Poisson-Boltzmann equation with excess salt present [26].

CONCLUSION

It has been shown using a modified version of the mean field theory of Miklavic Marcelja [1] for polyelectrolyte polymer brushes, which uses the non-linear Poisson-Boltzmann equation, that it should be possible for osmotic pressure due to the counterions to support a reasonably large load (about 10^5 Pa) with the brushes held sufficiently far apart to prevent entanglement of polymers belonging to the two brushes, which has been argued to account for most of the friction [19]. This load carrying ability is argued to persist in the presence of an amount of added salt that can be comparable to that found in living matter. Significant additional salt, however, provides screening which reduces the load carrying ability of polyelectrolyte brushes. The treatment provided in this article is applicable to polyelectrolyte brushes which either have a very small charge or are highly compressed, and not to brushes whose polymers are stretched to almost their maximum possible length. [27].

ACKNOWLEDGMENTS

I wish to thank the Department of Energy (Grant DE-FG02-96ER45585) for partial support of this work. I would also like to thank R. Tadmor for many discussions of his work and P. L. Hansen of MEMPHYS for insightful discussions.

REFERENCES

1. S. J. Miklavic and S. Marcelja, Journal of Physical Chemistry 92, 6718-6722 (1988).
2. J. Klein, Annual Reviews of Material Science 26, 581-612 (1996); H. Taunton, C. Toprakcioglu, L. J. Fetters, and J. Klein, Macromolecules 23,571-580 (1990); J. Klein, E. Kumacheva, D. Mahalu, D, Perahia, L. J. Fetters, Nature 370, 634 (1994); J. Klein, D. Perahia, S. Warburg, L. J. Fetters, Nature 352, 143 (1991); J. Klein, Colloids Surf. A 86, 63 (1994); J. Klein, E. Kumacheva, D. Perahia, D. Mahalu, S. Warburg, Faraday Discussions 98, 173 (1991); S. Granick, A. L. Demirel, L. L. Cai, J. Peanasky, Israel Journal of Chemistry 35, 75 (1995); J. Klein, Annu. Rev. Mater. Sci. 26, 581-612 (1996); J. Klein, Proc. Inst. Mech. Eng.: Part J: J. Engineering Tribology, Special Issue on Biolubrication 220, 691-710 (2006).
3. C. W. McCutchen, Wear 5, 1-17 (1962); D. Dowson, V. Wright and M. D. Longfeld, Biomedical Engineering 4, 160-165 (1969).
4. I. M. Schartz, and B. A. Hills, British Journal of Rheumatology 37, 21-26 (1998).
5. D. Dowson, Modes of lubrication of human joints. Symposium on Lubrication and wear in living and artificial joints, Inst. Mech, Eng., London 45-54 (1967).
6. C. W. McCutchen, Nature 184, 1284-6 (1959).
7. A. Maroudas, Proc. Inst. Mech. Eng. pp. 122-124 (1967).
8. V. C. Mow, and W. M. Lai, SIAM Rev. 22, 275-317 (1980).
9. T. Murukami, (JSME Int. J., Series).
10. G. A. Ateshian, J. Biomech. Eng. 119, 81-86 (1997).
11. T. B. Kirk, A. S. Wilson and G. W. Stachowiak, J. Orthop. Rheum. 6, 21-28 (1993).
12. C. W. McCutchen, Fed. Proc. 25, 1061-1068 (1996).
13. D. Swann, H. S. Slayter, and F. H. Silver, J. Biological Chemistry 256, 5921-5925 (1998).
14. G. D. Jay, Conn. Tiss. Res. 28, 71-78 (1992).
15. B. L. Schumacher, J. A. Block, T. A. Schmidt, M. B. Aydelotte, and K. E. Kuettner, Arch. Biochem. Biophys. 311, 144-152 (1994).
16. U. Raviv, S. Glasson, N. Kampf, J. F. Gohy, R. Jerome, J. Klein, Nature 425, 163 (2003).
17. J. Klein and E. Kumacheva, Journal of Chemical Phys. 108, 6996-7009 (1998); E. Kumacheva and J. Klein, Journal of Chemical Physics 108, 7010-7022 (1998).

18. J. Klein, E. Kumacheva, D. Mahalu, D. Perahla and L. J. Fetters, Nature 370, 634-636 (1994); H. J. Taunton, C. Toprakcioglu, L. J. Fetters and J. Klein, Macromolecules 23, 571-580 (1990).

19. J. B. Sokoloff, Macromolecules 40, 4053-4058 (2007).

20. S. T. Milner, T. A. Witten and M. E. Cates, Macromolecules 21, 2610-2619 (1988); T. A. Witten, L. Leibler, L. and P. A. Pincus, Macromolecules 23, 824-829 (1990); E. B. Zulina, V. A. Priamitsyn, O. V. Borisov, S. Vysokomol, Ser. A 30, 1615 (1989); Semenov, Sov. Phys. JETP 61, 733 (1985); Milner, S. T. Polymer Brushes. Science 251, 905-914 (1991); P. G. de Gennes, Advances in Colloid and Interface Science 27, 189-209 (1987); S. Alexander, J. Physique 38, 983 (1977); J. Klein, J. Phys.: Condens. Matter 12, A19 - A27 (2000); S. T. Milner, Science 251, 905-914 (1991); C. M. Wijmans, E. B. Zhulina and G. J. Fleer, Macromolecules 27, 3238-3248 (1994); J. Klein and P. Pincus, Macromolecules 15(4), 1129-1135 (1982).

21. R. Tadmor, J. Janik, J. Klein and L. J. Fetters, Physical Review letters 91, 115503 (2003).

22. J. U. Kim, M. W. Matsen, Euro. Phys. J. 23, 135 (2007); M. W. Matsen, J. Chem. Phys. 121, 1939-1948 (2004); M. W. Matsen, J. Chem. Phys. 117, 2351 (2002); S. T. Milner, T. A. Witten, Macromolecules 25, 5495-5503 (1992); R. R. Netz and M. Schick, Macromolecules 31, 5105-5122 (1998); T. Kreer, S. Metzger, M. Muller, K. Binder, J. Baschnagel, J. Chem. Phys. 120, 4012-4022 (2004).

23. G. S. Grest, Advances in Polymer Science 138, 149 (1999); M. Murat, G. S. Grest, Macromolecules 22, 4054-4059 (1989); G. S. Grest, J. Chem. Phys. 105,5532 (1996); G. S. Grest, Current Opinion Colloid Interface Science 2, 271 (1997); G. S. Grest, in "Dynamics in Small Confined Systems III (Materials Research Society, Pittsburg, 1997), ed. J. M. Drake, J. Klafter, R. Kopelman, vol. 464, 71; G. S. Grest, Phys. Rev. Lett. 76, 4979 (1996).

24. S. A. Safran , "Statistical Thermodynamics of Surfaces, Interfaces and Membranes," (Addison-Wesley, Reading, MA,1994), pp. 158-159.

25. J. B. Sokoloff, "Theory of Lubrication due to Poly-Electrolyte Polymer Brushes," Arxiv preprint archives cond-mat 0706.1252.

26. P. Pincus, Macromolecules 24, 2912 (1991); M. N. Tamashiro, E. Hernandez-Zapata, P. A. Schorr, M. Balastre, M. Tirrell and P Pincus, J. Chem. Phys. 115, 1960 (2001).

27. E. B. Zhulina and O. V. Borisov, J. Chem. Phys. 107, 5952-5966 (1997); A. Naji, R. R. Netz and C. Seidel, Europhysical Journal E 12, 223-237 (2003).

Mater. Res. Soc. Symp. Proc. Vol. 1049 © 2008 Materials Research Society 1049-AA05-15

Tip-Induced Calcite Single Crystal Nanowear

Ramakrishna Gunda, and Alex A. Volinsky
Department of Mechanical Engineering, University of South Florida, 4202 E. Fowler Ave. ENB118, Tampa, FL, 33620

ABSTRACT

Wear behavior of freshly cleaved single crystal calcite ($CaCO_3$) was investigated by continuous scanning using the Hysitron Triboindenter in ambient environment as a function of scanning frequency (1 Hz – 3 Hz) and contact load (2 µN – 8 µN). At lower loads below 4 µN, initiation of the ripples takes place at the bottom of the surface slope, which continue to propagate up the slope as scanning progresses. The orientation of these ripple structures is perpendicular to the long scan direction. As the number of scans increases, ripples become fully developed, and their height and periodicity increase with the number of scans. At 6 µN normal load, tip-induced wear occurs as the tip begins removing the ripple structures with increased number of scan cycles. As the contact load increased further, ripples did not initiate and only tip-induced wear occurred on the surface, and saturated after 20 scans. At 1 Hz frequency wear takes place as material slides towards the scan edges when the tip moves back and forth. Material removal rate increased with contact load and it is observed that the number of scans required to create a new surface is inversely proportional to the contact load. Possible mechanisms responsible for the formation of ripples at higher frequencies are attributed to the slope of the surface, piezo hysteresis, system dynamics, or a combination of effects. The wear regime is due to abrasive wear. Single crystal calcite hardness of 2.8±0.3 GPa and elastic modulus of 75±4.9 GPa were measured using nanoindentation and used to determine the wear mode.

INTRODUCTION

The development of the nano-mechanics field over the past few decades produced a significant amount of methods for determination of mechanical and tribological properties of materials at micro and nano scales. Among others, nanoindentation and micro/nanotribology methods have been considerably developed, including depth sensing nanoindentation and Atomic Force Microscopy (AFM)/Scanning Tunneling Microscopy (STM). These are powerful and versatile tools for surface topography characterization and local mechanical properties measurements at small scales. In addition, it is possible to utilize the scanning nanoindenter capable of modifying the materials structure. As the use of coatings constantly increases in the field of nanotechnology, it is important to know the behavior of materials at small scales. In recent years the formation and characterization of nanometer-sized structures have attracted a great deal of interest.

One of the nanotechnology research motivations is the construction of nanosized surface patterns. In a typical wear experiment a hard material is scanned over the tested material surface, resulting in a wear rate measurement in terms of the removed material depth or volume as a function of normal applied load and the number of wear cycles. When applying moderate forces in combination with repeated scanning over the same region, one could observe a periodic ridged pattern, called wear ripples, oriented perpendicular to the tip motion direction. These nanopatterns or ripple structures can also be produced by other methods such as erosion of materials by ion sputtering and by abrasive particles bombardment. Under off-normal ion-beam incidence, a periodic height modulation in the form of ripples or wavelike structures with submicron wavelength develops during low-energy ion bombardment of single crystalline Si (100) [1], single crystalline metals [2] and glasses [3]. Recently, Krok *et.al.* found that ion bombardment of InSb at an oblique angle of incidence led to the formation of wire-like structures on the surface with a diameter of tens of nanometers [4]. Tip-induced nanopatterns or nanowear ripples on the material surface can be produced by scanning it repeatedly for a number of cycles and applying moderate normal forces using AFM or scanning Triboindenter. Although the main AFM function is imaging surfaces, it can also scan them continuously and modify their structure. The Triboindenter is mainly used for measuring hardness and elastic modulus at the nanoscale. Using the Triboindenter scanning feature, one can produce nanowear ripples or nanostructures on the material's surface. The interaction between the tip of the scanning probe apparatus (AFM or the Triboindenter) and tested surface is a complex process, which depends on material properties and the scanning parameters. By varying the applied force and scanning speed, tested material surfaces can exhibit different regimes of wear caused by the tip.

Initially tip-induced wear patterns were observed in InSb crystals using atomic force microscope in the ultra high vacuum environment (UHV-AFM) [5]. Few tens of nanonewtons of normal load were applied to scan repeatedly over the $1x1$ μm^2 InSb (100) surface, which resulted in a creation of ripples perpendicular to the scan direction. Ripple patterns were not observed when repeated scanning was performed in the ambient environment with AFM. Similar experiments have been conducted on KBr and Al single crystals in the ambient environment using a nanoindenter [6]. Ripples were observed on KBr after 20 scan cycles at 3 Hz scan frequency over a $5x5$ μm^2 surface area with a 2 μN normal force. The height of the ripples was 100 nm, and they were spaced 1 μm apart. Tip-induced wear ripples were also reproduced on gold samples in the presence of water using the Hysitron Triboindenter [7].

In this work, we made an effort to produce tip-induced nanowear ripples on calcite single crystals using a diamond Berkovich tip of the Hysitron Triboindenter in ambient environment. The objective was to replicate the nanowear ripples observed in the UHV-AFM and to determine the mechanism responsible for the formation of these ripples.

EXPERIMENTAL DETAILS

Freshly cleaved natural calcite single crystal was scanned repeatedly in the ambient environment using a Hysitron Triboindenter. This tool has the ability to continuously scan the material surface, while producing surface topography images. The piezo scanner of the Triboindenter can scan $80x80$ μm^2 or less square area using a rigidly fixed diamond tip. Tip-induced wear patterns were observed as a function of scanning frequency (1 - 3 Hz) and normal load (2 - 8 μN). Scanning speed of the tip depends on the frequency and the scan size. For $20x20$ μm^2 scan size, the velocity of the tip will be 120 $\mu m/sec$ at 3 Hz scanning frequency. Tip-induced wear was achieved by repeated scanning of $20x20$ μm^2 areas of the calcite surface at varying loads and frequencies. At 3 Hz scanning frequency wear ripples started initiating during the first scan and propagated as the number of scans increased. Similar results were obtained on KBr and Al single crystals [6] in the ambient environment using the same system as well as using UHV-AFM on InSb semiconductor surface [5]. As the contact load increases beyond 6 μN, material started moving to the scan edges and ripple structures were never initiated. Similarly, at 1 Hz scanning frequency, a ripple structure was not initiated and wearing of the surface layer took place.

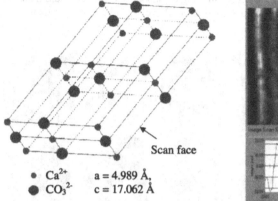

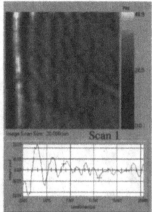

Figure 1. Calcite crystal structure. Adapted with permission from "Calcite: structure". Encyclopedia Britannica Online.

Figure 2. Topography of the calcite surface at 3 Hz frequency and 2 μN normal load and its profile taken at the center.

Calcite has a chemical formula of $CaCO_3$, and is one of the most widely distributed minerals on the Earth. Figure 1 shows the true unit cell ($2[CaCO_3]$) of the acute rhombohedron calcite crystal structure.

Figure 2 shows the topography image and the line profile of the calcite single crystal after the first scan. Here, the surface horizontal slope is ~200 nm over a 20 μm scan length. Ripples initiated at the bottom of the surface slope during the first scan. As scanning continued, ripples propagated towards the other end of the scan edge, and the ripples orientation was perpendicular to the long scan direction.

RESULTS AND DISCUSSION

Figure 3 shows a series of calcite topography images after repeated scanning at 3 Hz frequency, 2 μN and 6 μN normal loads at the intermittent scans. At 2 μN normal load, ripples started initiating during the first scan and propagated as scanning progressed. One could see ripples on the surface with a periodicity of 2.3 μm after 10 scan cycles. They merged together after 10 scans and the rms surface roughness increased as the ripples height increased from 30 to 145 nm at the end of 130 cycles. Even after 130 scans, ripples with higher periodicity can be seen and the top layer of the surface was not completely removed. At 6 μN, no ripples were initiated during the first scan, while they started to appear after five scans at the bottom of the surface horizontal slope and propagated towards the other scan edge as scanning continued. The initial ripple height was 55 nm and as scanning continued, the ripples height decreased to 42 nm due to surface wear. As the number of scans increased, the tip started removing the ripples on one end, while propagating them at the other end of the scan edge. One can clearly see material removal from the surface as the scanning progressed with ripples periodicity decreased from 3.4

to 2.5 μm. After 28 scans, next surface layer of the calcite appeared and all the ripples were swiped away to the scan edges. When normal load increased above 6 μN, ripples were not initiated on the crystal surface and the tip started removing the calcite surface layer, and material wear took place. A trench, 10-15 nm deep is observed after 30 cycles, which corresponds to about 10 top unit cells of calcite removed.

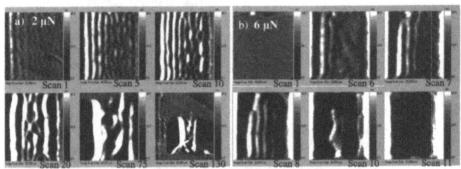

Figure 3. Topography images of calcite after repeated scanning, showing ripple initiation and propagation as scanning progressed at a) 2 μN and b) 6 μN normal load.

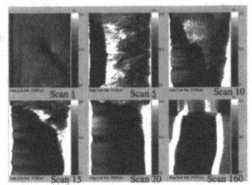

Figure 4. Topography images of calcite after repeated scanning at 1 Hz scanning frequency and 2 μN normal load.

Figure 4 shows series of topography images of the calcite surface at 1 Hz scanning frequency and at 2 μN normal load. In this case, calcite surface topography has a horizontal slope of 280 nm over 20 μm. Ripple structures were not initiated at these scanning parameters and as the scanning continued, the tip started removing the material from the bottom of the slope instead of forming ripples on the surface. The top surface calcite layer was completely removed after 20 scans without forming the ripples.

A trench, 15 nm deep can be seen after 20 scans at this frequency and normal load. As the load increased to 6 μN, the tip started removing the calcite surface at a faster rate and no ripples were initiated. A new surface layer could be seen with less number of scans when compared to the previous case and a trench of 15 nm deep can be seen after 10 scan cycles. From the initial surface topography images, it is clear that the surfaces had a horizontal slope in the range between 100 nm and 200 nm for the 20 μm scan length. It was observed that in all cases ripples initiated and started propagating at the bottom of the slope, i.e. at the left side of the scan edge, which is attributed to the sample tilt. Similar behavior was seen in KBr in the ambient environment using Hysitron Triboindenter [6] and on the InSb semiconductor surface in the

UHV-AFM [5]. A possible mechanism for the initiation of ripples is due to piezo hysteresis and surface slope. While the indenter is moving up along the surface slope, the piezo will work against gravity. Since the indenter scans the surface with constant velocity, the piezo takes little time to respond when indenter is in transition motion from forward to backward direction and vice-versa. This response time in the piezo could cause digging effect of the tip into the sample, which causes initiation of the ripple structure. As the ripple pattern behavior was observed at a relatively high frequency of 3 Hz, and removal of surface material occurred at a lower frequency of 1 Hz, the self-excited vibration of the tip could have caused the ripples to propagate along the surface of the crystal. This phenomenon is equivalent to chatter observed at the macroscopic scale [8, 9].

Figure 5 shows the wear behavior of the calcite single crystal at 1 Hz frequency and 2 µN normal load. Tip-induced wear in the negative z direction linearly increased for 6 cycles, removing approximately 15 calcite unit cells, and then saturated as scanning progressed further. This might be attributed to the hardening of the surface or dislocation density increasing during scanning. Tip-initiated material removal along the scan edge from the bottom slope increased as scanning continued. This lateral wear removal is approximately linear with the number of scans.

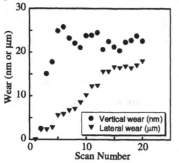

Figure 5. Wear behavior of the calcite single crystal at 1 Hz frequency and 2 µN load.

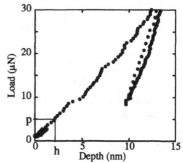

Figure 6. Load-displacement curve of calcite single crystal showing contact depth determination.

Single crystal calcite hardness and elastic modulus were measured using the Hysitron Triboindenter with the diamond tip. In this measurement, partial loading and unloading cycle was used to evaluate these properties as a function of contact depth. Figure 6 shows part of the partial loading and unloading load-displacement curve. The contact pressure under the tip surface was measured from the normal load applied while scanning over the corresponding contact area. Contact area was calculated from the geometry of the tip and contact depth corresponding to the normal loads. Figure 7 shows the hardness (2.8±0.3 GPa) of the calcite single crystal as a function of the contact depth and the values corresponding to the horizontal lines show the contact pressures applied by the tip during the scanning at various normal loads. At small loads, it is evident that contact pressure is greater than the hardness of the calcite single crystal and tip-induced plastic wear of the material takes place during scanning. Figure 8 shows the calcite single crystal elastic modulus of 75±4.9 GPa, which does not vary significantly with the contact depth.

CONCLUSIONS

Formation of tip-induced nanowear ripples was studied on the surface of single crystal calcite in the ambient environment using a Hysitron Triboindenter. Ripple structures were observed at different contact loads and at different frequencies by scanning the surface repeatedly with a Berkovich diamond tip. At 3 Hz scanning frequency, ripples initiated at the bottom slope of the scanned surface and propagated as the number of scans increased at lower loads below 6 μN. At 1 Hz scanning frequency, only material removal on the surface was observed and ripples did not initiate. Similar experiments were also conducted on InSb, Si, GaAs, and Quartz single crystals, and polycrystalline Bismuth. It was observed that nanoripples did not initiate in these materials.

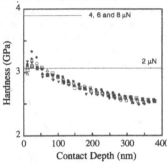

Figure 7. Calcite hardness and contact pressure under varying normal loads.

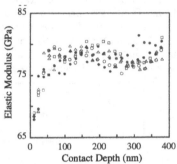

Figure 8. Elastic modulus of the calcite single crystal.

ACKNOWLEDGMENTS

This work was supported by the National Science Foundation (CMMI-0631526 and CMMI-0600231), and the NSF IREE supplement. The authors would like to thank Prof. Szymonski's group at Jagiellonian University in Krakow, Poland, and Hysitron Inc. for support.

REFERENCES

1. G. Carter and V. Vishnyakov, *Phys. Rev. B* 54, 17 (1996) 647
2. S. Rusponi, G. Costantini, C. Boragno, and U. Valbusa, *Phys. Rev. Lett.* 81, (1998) 4184
3. P. F. A. Alkemade, *Phys. Rev. Lett.* (2006) 96, 107602
4. F. Krok, J.J. Kolodziej, B. Such, P. Piatkowski, M. Szymonski, *Appl. Surf. Sci.* 210 (2003) 112
5. M. Szymonski, F. Krok, B. Such, P. Piatkowski, J.J. Kolodziej, *3rd EFS Nanotribology Conference,* Lisbon, Portugal, 18-22 September, 2004
6. M. Pendergast, A.A. Volinsky, *Mat. Res. Soc. Proc.* 1021 E (2007)
7. Xiaolu Pang, A.A. Volinsky, Kewei Gao, submitted to Journal of Materials Research
8. P. Zelinski, Modern machine shop magazine, October 2005
9. E.P. Degarmo, J.T. Black, R.A. Kohser, John Wiley & sons, 9th ed., (2003) 504

Mater. Res. Soc. Symp. Proc. Vol. 1049 © 2008 Materials Research Society

Study on the local damage mechanisms in WC-Co hard metals during scratch testing

Siphilisiwe Ndlovu, Karsten Durst, Heinz Werner Hoeppel, and Mathias Goeken
Department of Materials Science and Engineering, University of Erlangen-Nuernberg,
Erlangen, 91058, Germany

ABSTRACT

The effect of the cobalt content and WC grain size on the deformation behaviour of WC-Co hard metals was investigated by studying materials with a varying WC grain size and cobalt content. The WC grain size ranged from 0.25 to 2.65 µm and the binder content ranged from 6 to 15 wt%. Single and multiple scratch tests were conducted using a Nanoindenter with a Berkovich diamond tip and the load ranged from 5 to 500 mN with a tip sliding velocity of 10 µm/s. Several damage mechanisms were observed and these show a combination of ductile and brittle wear. The bulk properties i.e. composite properties of the hard metal determine the wear in the 6 wt% Co samples. On the other hand the 15 wt% Co samples exhibited a localised response to the wear i.e. the wear is determined by the individual phases in the hard metal.

INTRODUCTION

WC-Co hard metals are well established powder metallurgy products. The unique composite structure of hard WC grains in a tough cobalt matrix results in excellent wear resistance. Recent developments have led to the production of nanostructured WC-Co hard metals which consist of nanoscale tungsten carbide grains in a cobalt matrix. These nanostructured hard metals are reported to have enhanced wear properties as a result of their increased hardness and the increased constraint of the WC grains in the binder phase which results from the reduced binder mean free path [1]. Nanoindentation is a well established experimental technique to investigate the localised mechanical properties of a system and a Nanoindenter can also be used for scratch testing of materials [2]. The local wear behaviour of WC-Co hard metals will thus be investigated using scratch testing. The mechanical and wear properties of WC-Co hard metals on the local scale will be investigated with emphasis on the WC grain size and the cobalt content.

EXPERIMENT

Materials

Seven commercial hard metal grades with varying cobalt content ranging from 6 to 15 % were tested in this study. The grades are classified as ultra fine (UFG), medium (MG) and coarse grained (CG) depending on the size of the WC grains. The characterisation data is given in Table 1.

Table I: Composition and properties of the investigated WC-Co cemented carbides

Sample	Average WC grain size (μm)	Description	Binder wt%	Vickers hardness H_v	Indentation hardness (GPa)	Young's Modulus (GPa)
J15	2.65	CG	15	1017	12.7	555
NY15	0.25	UFG	15	1486	19.7	561
T06M	1.21	CG	6	1380	18.4	530
T06MF	0.60	MG	6.5	1575	22.1	638
T06MG	0.48	UFG	6	1760	21.8	476
T06F	0.66	MG	6	1710	22.0	615
T06SMG	0.25	UFG	6	1940	25.5	623

Instrumented scratch testing

Nanoscratch tests on the hard metals were performed using a Nano Indenter XP with a load controlled head. The maximum load capacity for the standard system is 500 mN with a precision of less than 1 mN. All the scratch tests were performed with one corner of the Berkovich indenter in the scratch direction since the tip orientation has been found to have an effect on the wear behaviour [3, 4]. The normal load ranged from 5 mN to 500 mN with a tip velocity of 10 μm/s. Tip calibration was performed using fused silica before each series of tests to monitor the tip function and no significant change in the tip shape was observed. The worn samples were examined using an atomic force microscope (AFM) and a scanning electron microscope (SEM).

RESULTS

Low cobalt content

The scratch depth was determined using the section analysis mode of the AFM. Figure 1 shows the increase in scratch depth with increasing load for the samples containing 6 wt% cobalt binder. A plot of the grain size versus the scratch depth was made at a load of 100 mN (figure 2) and shows that an increase in the WC grain size leads to an increase in the scratch depth i.e. a decrease in scratch resistance. A lower scratch depth was measured for the UFG sample (T06SMG) compared to the coarse grained hard metal (T06M). At a load of 100 mN the coarse grained sample exhibited a scratch depth of 297 nm compared to 227 nm for the UFG sample (T06SMG) for a single scratch.

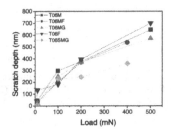 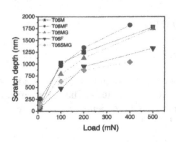

(a)   (b)

Figure 1: Plot of scratch depth against scratch load for the 6 wt% Co samples
(a) single scratch and (b) multiple scratch test (20 passes)

The scratch depth was found to increase as expected with multiple scratches. The scratch depth increased from 298 nm to 1.022 µm and from 227 nm to 632 nm for T06M and T06SMG respectively, after 20 scratch passes. Scratch testing at low loads resulted in grooving of the WC grains and the formation of cracks in the WC grains. Multiple scratches at low loads led to the removal of material via WC grain fracture and fall out. In the UFG sample a film was observed on the surface of the damaged region as discussed in earlier work [5].

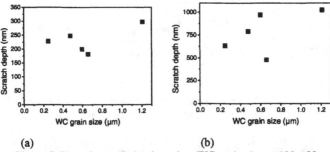

(a)   (b)

Figure 2: Plot of scratch depth against WC grain size at 100 mN
(a) single scratch and (b) multiple scratch test

At high loads single scratches resulted in more severe WC grain cracking and ductile deformation of the WC grains was also observed with extensive slip line formation as shown in figure 3a. The deformation of the cobalt binder was also observed, it attains a porous-like structure during scratching which makes it easier for WC grain fall out (figure 3b). Multiple scratches at high load led to the formation of a mechanically mixed layer composed of WC fragments and cobalt binder. There was also material loss as a result of grain fall out and fracture.

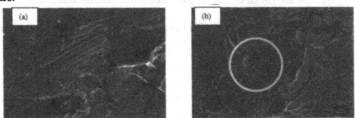

Figure 3: Micrographs of (a) T06F and (b) T06M after a single scratch test at 500 mN

High cobalt content

The UFG sample (NY15) displays a consistently higher scratch depth at all tested loads as shown in figure 4. Multiple scratch tests saw a significant increase in scratch depth for both samples with the UFG sample exhibiting a poorer behaviour. For example the change in depth from a single to a multiple scratch test at a load of 500 mN is 150 % for the coarse grained material and 84 % for the UFG sample.

An examination of the sample with a SEM showed that the wear mechanisms in the UFG sample were much more severe. Grooving of the WC grains and pile-up of the binder phase was observed in the CG sample after single scratch tests at low loads and multiple scratch testing resulted in chipping of the WC grains due to crack growth and intersection. Extrusion of the Co binder was also observed (figure 5a). The UFG sample exhibited cracking and chipping of the WC grains and fall out of the smaller WC grains as shown in figure 5b. Multiple pass tests led to increased material removal and the formation of a mechanically mixed layer on the material surface.

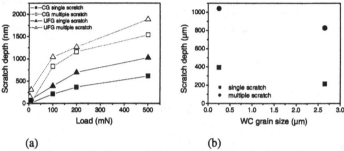

(a)   (b)

Figure 4: (a) The variation in scratch depth with load for the 15 wt% Co samples (b) scratch depth plotted against WC grain size at 100 mN

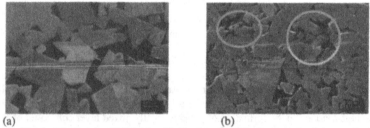

(a)   (b)

Figure 5: SEM micrograph of WC-15 wt% Co sample after single scratch test at 5 mN (a) CG sample and (b) UFG sample (please note the different scale bars)

The typical wear damage observed in the CG sample after single scratch tests at higher loads is shown in figure 6a. This shows the formation of slip lines in the larger WC grains and crack formation approximately normal to the slip line direction. Cutting and chipping of the WC grains was also found which led to the removal of material. Multiple scratch tests resulted in more material loss and the formation of a mechanically mixed layer. Single and multiple scratching in the UFG sample at high loads resulted in more extensive grain removal, fracture and the formation of the mechanically mixed layer which was also observed at multiple passes at low loads. The marked area in figure 6b shows a region on the edge of the scratch where WC grains have been removed from the bulk material after a single scratch with a 100 mN load.

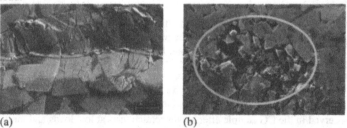

(a)   (b)

Figure 6: WC-15 wt% Co after a single at 100 mN (a) CG sample (b) UFG sample

T06SMG and NY15 have a WC grain size of 250 nm and were compared to determine the influence of binder content. The 6 wt% Co sample showed a lower scratch depth than the 15 wt % Co sample for both single and multiple scratch tests

DISCUSSION

Low cobalt content

During a scratch test a load is first applied and the indenter penetrates the sample and then moves across the sample at the chosen velocity. At low loads the penetration depth of the indenter is very low and it only scratches the surface of the material which leads to cracks and grooves in the WC grains. Multiple scratching at low loads leads to fracture of the WC grains as a result of crack growth and intersection. The fractured grain segments may remain embedded in the material or be lost. Grain fall out occurs due to the movement of the sharp indenter tip which repeatedly pushes against grains which are loosely anchored as a result of binder extrusion. Grain fall out in the UFG samples was very limited, this is because these materials are harder and therefore the penetration depth of the indenter is lower and the damage mainly occurs on the surface of the material. At higher loads the penetration depth is increased and the damage occurs deeper in the material surface. A cross section of a scratch is shown in figure 7a and this shows that the cracks in the grains extend into the bulk material. The micrograph also shows that the material is not completely pore free and small pores can be seen in the material as marked. The presence of pores can be detrimental to the material during scratch testing as the pores compromise the material integrity.
The binder phase exhibits a porous like structure after scratch testing, which makes it much easier for the WC grains to be removed from the material. This explains the extensive grain fall out observed in the WC-6wt%Co samples during multiple scratch testing at high loads.

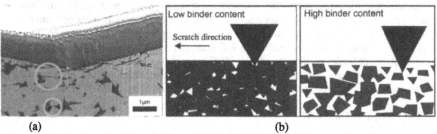

(a)   (b)

Figure 7:(a) Cross-section of multiple scratch on T06MF (b) Schematic of wear mechanism

High cobalt content

At low scratch loads the coarse grained material exhibits very low scratch depths and the damage is confined to grooving of the WC grains and extrusion of the cobalt binder. The UFG material on the other hand shows much more severe damage mechanisms. Single scratches resulted in grooving, cracking and chipping of the WC grains and also the fall out of WC particles in the UFG sample. A schematic of the proposed wear mechanism is shown in figure 7b. A low load results in a relatively low penetration depth, however due to the high content of the binder phase the penetration depth is higher than that in the WC-6Wt%Co samples. In the CG samples the grains are very large and even at the higher penetration depths only the surface of the WC grains is damaged. In comparison the grain size in the UFG sample is comparable to the penetration depth and the movement of the indenter across the material can lead to the removal of the WC grains. At higher loads a similar mechanism would be expected. At a load of 500 mN a scratch depth of 614 nm was observed for the CG material after a single scratch test, this is very small in comparison to the average grain size of 2.65 μm. The scratch depth for the UFG sample at the same load was 1.024 μm which is more than four times the average grain size of the material.

Thus the material behaviour in the WC-15wt%Co samples during scratch testing is influenced by the local material properties and not the bulk material properties. It must however be noted that multiple scratch tests at high loads led to similar wear mechanisms in both the UFG and CG 15 wt% samples. In the case of the WC-6wt%Co samples the bulk material properties appear to play a bigger role in the material behaviour regardless of the WC grain size. This phenomena was reported for the erosive wear of WC-Co hard metals by Anand and Conrad [6]. They found that fine-grained materials responded in bulk to erosive attack when the damage zone was comparable to the microstructural dimensions. In coarser materials the microstructure was comparable to the damage zone and the constituent phases of the material responded individually to the erosive attack.

The penetration depth of the indenter during scratch testing is important in determining the wear that takes place. When the penetration depth is high the indenter is able to push out the smaller grains from the bulk material more easily. When the binder content is low the indenter encounters the hard phase more often and the penetration depth is low confining the damage to the surface of the material. However, with a high cobalt content the soft binder phase is more abundant and the penetration depth of the indenter is generally higher and smaller grains are more easily removed.

CONCLUSIONS

The WC-6 wt% Co hard metals investigated were found to respond in bulk to the scratch testing conducted whereas the WC-15wt%Co samples were influenced by the local material properties. A finer WC grain size was found to result in a better wear performance for the 6 wt% Co sample and a coarser grain size was better for the 15 wt% samples. The UFG 15 wt% Co sample showed extensive WC grain fall out, not observed in the CG material. A lower binder content (for a given WC grain size) was found to improve the scratch resistance of the material. Therefore it could also be included that ideal microstructure for good scratch resistance would be a low binder content and ultra-fine WC grain size.

ACKNOWLEDGEMENTS

The authors wish to thank Tigra Hartstoff GmbH for the supply of materials. The German Academic Exchange Service (DAAD) is gratefully acknowledged for its financial support.

REFERENCES

1. K. Jia, T.E. Fischer, Sliding wear of conventional and nanostructured cemented carbides, Wear 203/204, 310-318 (1997)
2. K. Durst, M.Göken, Micromechanical characterisation of the influence of Rhenium on the mechanical properties in nickel-base superalloys, Materials Science and Engineering A 387-389, 312-316 (2004)
3. S. W. Youn, C. G. Kang, Effect of nanoscratch conditions on both deformation behaviour and wet-etching characteristics of silicon (100) surface, Wear 261, 328-337 (2006)

4. D. Mulliah, D. Christopher, S. D. Kenny, Roger Smith, Nanoscratching of silver (100) with a diamond tip, Nuclear Instruments and Methods in Physics Research B202, 294-299 (2003)
5. S. Ndlovu, K. Durst, M. Göken, Investigation of the wear properties of WC-Co hard metals using nanoscratch testing, Wear 263, 1602-1609 (2007)
6. K. Anand and H. Conrad, Microstructue and scaling effects in the damage of WC-Co alloys by sinfle impacts of hard particles, Journal of Materials Science 23, 2931-2942 (1988)

Modeling, Simulation and
Analysis of Indentation Data

Mater. Res. Soc. Symp. Proc. Vol. 1049 © 2008 Materials Research Society   1049-AA06-03

Mapping of the Initial Volume at the Onset of Plasticity in Nanoindentation

T.T. Zhu[1], K.M.Y. P'ng[1], M. Hopkinson[2], A.J. Bushby[1], and D.J. Dunstan[1]

[1]Centre for Materials Research, Queen Mary University of London, London, E1 4NS, United Kingdom

[2]Department of Electronic and Electric Engineering, University of Sheffield, Sheffield, S1 3JD, United Kingdom

ABSTRACT

Understanding the finite volume throughout which plastic deformation begins is necessary to understand the mechanics of small-scale deformation. In indentation using spherical indenters, conventional yield criteria predict that yield starts at a point on the axis and at a depth of half the contact radius. However, Jayaweera *et al* (Proc. Roy. Soc. 2003) [4]concluded that yield occurs over a finite volume at least 100 nm thick. Semiconductor superlattice structures, in which the stress and thickness of individual layers can be varied and in which known internal stresses can be incorporated, open up new possibilities for investigation that cannot be achieved by varying external stresses on a homogenous specimen. We have designed samples with bands of highly strained InGaAs superlattice, which are essentially bands of low yield-stress material devoid of other metallurgical artifacts. These bands are placed at different depths in a series of samples. Spherical indenters with a range of radii were used to determine the elastic-plastic transition. The stress field from different sized indenters interacts with the low yield-stress material at different depths below the surface to map out the size of the initial yield volume.

INTRODUCTION

Small-scale mechanical behaviour is at the cutting-edge of research in materials science and applied mechanics. Plastic yield strength is one of the most fundamental subjects in materials science and engineering field. Also, it is a very useful in materials designing and applications, since it certainly helps to predict the reliability of materials. The plastic yield strength is not only dependent on the material structure but also on the loading configuration. Nanoindentation testing is an excellent, usually non-destructive method for testing the mechanical properties of materials at small scales. The yield behaviour in nanoindentation is a very popular and attractive topic nowadays. [1-4]

With spherical indenters, the stress fields can be obtained using Hertzian contact mechanics and yield is predicted to initiate at the point where a suitable some yield criterion is first exceeded. This point is on the indenter axis and at a depth of half the contact radius. However, on the basis indirect experimental evidence, Jayaweera *et al.* [4] proposed that yield in spherical nanoindentation should initiate throughout a finite volume at least 100nm across, perhaps as large as a micron. In this paper, a unique experimental method is demonstrated to observe directly the real initial volume of yield for spherical indentation. Results indicate that the size of the initial volume varies for different radius indenters and also indicate where plasticity initiates.

EXPERIMENTAL MATERIALS AND METHODS

Materials design idea and structure

It has been demonstrated that yield strength of coherently strained $In_xGa_{1-x}As$ superlattice depends on the thickness and coherency strains of the layers of the superlattice. For alternating tensile and compressive layers, the relevant parameter is

given by [4]: $F = \dfrac{|\varepsilon_t h_t - \varepsilon_c h_c|}{h_t + h_c}$ where ε is strain, h is thickness, and the subscripts t

and c denote tensile and compressive respectively. Details of the F factor and the observed linear decrease of the yield strength with F are given elsewhere [4, 5]. Here, we use this property to build semiconductor structures with bands of superlattice with values of F greater than the surrounding material and hence with reduced yield strength.

Four structures were grown by Molecular Beam Epitaxy (MBE) on (001) InP substrate. MBE produces atomically flat thin strained layers with low dislocation densities determined by the substrate. Two control structures have constant F factor throughout; the first, with $F = 0$ throughout, is simply 2.5μm of $In_{0.53}Ga_{0.47}As$ (lattice matched) epilayer, while the second, with $F = 0.8$ throughout, is a superlattice of alternating 50nm $In_{0.648}Ga_{0.352}As$ and $In_{0.416}Ga_{0.884}As$ layers, with a total thickness 2.5μm. The other two structures have 300nm layers of $F = 0.8$ superlattice on top of or buried in inside $F = 0$ material, and are illustrated in Fig. 1. The superlattices consist of 6 layers each 50nm thick. In the structure B1, the superlattice is at the surface. In the structure B2, the superlattice is buried under 700nm of $F = 0$ material.

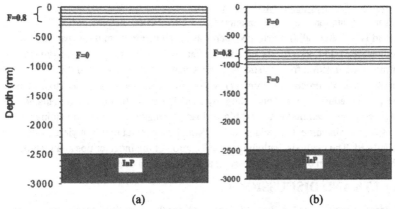

(a)   (b)

Figure 1. Schematics of the layer structures: (a) B1, 300nm of $F = 0.8$ superlattice is at the surface of 2.5μm of InGaAs grown on GaAs; (b) B2, the same except that the 300nm of $F=0.8$ superlattice is buried under 700nm of $F = 0$ InGaAs.

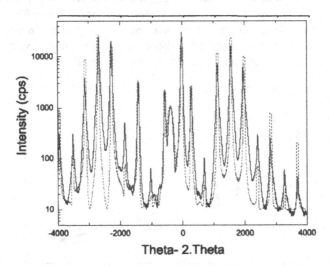

Figure 2. The experimental high-resolution X-ray diffraction rocking curve (solid line) and the simulated rocking curve (red broken line) for the control sample with $F = 0.8$.

The samples were characterised by high-resolution X-ray diffraction (HRXRD), which is a good tool for confirming that the structure grown is what was intended [6]. Fig. 2 shows a typical example of an HRXRD reciprocal space map for the control sample $F = 0.8$ projected as a rocking curve and compared with the simulated projection.

<u>Nanoindentation test using partial unloading</u>

Nanoindentation tests were conducted on a UMIS 2000 instrument (CSIRO, Lindfield NSW, Australia). Each specimen was tested using the multiple partial unloading method [7, 8] with six spherical indenters from 450nm to 20µm radius to determine the transition from elastic to plastic behaviour. In this method the indentation loading proceeds incrementally. Following each load increment to a force F_i, the load is reduced (partial unloading) by a small amount (in our experiments, to $0.75F_i$) before proceeding to the next higher load increment, F_{i+1}. After obtaining the entire stress-strain curve, the yield pressure P_Y and the contact radius at yield a_Y can be calculated. The methods, including the calibration of the indenter tips and methods for the determination of the yield pressure, are described in detail in [8, 9].

RESULTS AND DISCUSSIONS

Results are presented in Fig. 3. As expected, for the same indenter size or similar contact radius, the lattice-matched $F = 0$ control structure is stronger than the strained $F = 0.8$ superlattice control, in good agreement with Jayaweera (2003). Also, the size effect in which yield strength increases with decreasing indenter radius or contact radius is clearly observed. Crucially, the yield pressures in the structures B1 and B2 lie between the controls.   It is not our purpose here to investigate the size effect behavior. Consequently, we normalize the data of Fig. 3 to show more clearly how the superlattices influence yield pressure in the structures B1 and B2.

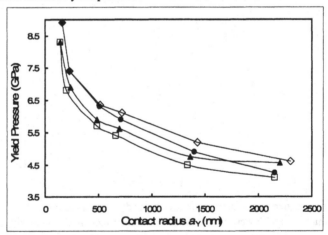

Figure 3. The yield pressure P_Y is plotted against the contact radius at yield a_Y for all four samples with various spherical indenters. Open diamonds: the $F = 0$ control sample; Open squares: the $F = 0.8$ control sample; Solid triangles: the superlattice layer sample B1; Solid circles: superlattice layer sample B2. Solid lines are guides for the eye.

Since these structures are mostly $F = 0$ material we may consider the highest yield pressure expected in them to be that of the $F = 0$ control and we normalize this to unity ($e = 1$). On the other hand, if yield starts in the $F = 0.8$ material we would expect to obtain the same yield pressure as in the $F = 0.8$ control. We normalize this to zero ($e = 0$). Then other values of yield pressure P_Y in the structures B1 and B2 may be described by the factor

$$e = \frac{P_Y - P_{0.8}}{P_0 - P_{0.8}}$$

where P_0 and $P_{0.8}$ denote the yield pressures for the control $F = 0$ and $F = 0.8$ samples. The quantity $1 - e$ might be described as the efficiency with which the 300nm superlattice layer reduces the yield pressure to the value of the $F = 0.8$ control sample. In Fig. 4 values of e are plotted against contact radius at yield, a_Y. By the definition of e, the data for the $F = 0$ control structure lie on the line $e = 1$; and the data for the $F = 0.8$ control lie on the x axis, $e = 0$, and therefore are not plotted in Fig 4. The data for B1 and B2 lie between 0 and 1.

Conventionally, it is supposed that the yield begins at the point of maximum shear stress, which is at a depth of half the contact radius, $\frac{1}{2}\,a_Y$, beneath the centre of the indentation. If this were true, in the structure B1, when $\frac{1}{2}\,a_Y < 300$ nm, yield would begin in the superlattice band and the yield pressure would be the same as in the $F = 0.8$ control; that is, e should be 0. When $\frac{1}{2}a_Y > 300$ nm, yield would begin below the superlattice band and the yield pressure would be the same as in the $F = 0$ control; that is, e should be 1. This behaviour is marked schematically in Fig.4 as the broken line. Similarly, in the structure B2, we would expect e to be zero only for

$700 < \frac{1}{2}a_Y < 1000$ nm, and unity elsewhere, as indicated schematically by the solid line in Fig.4.

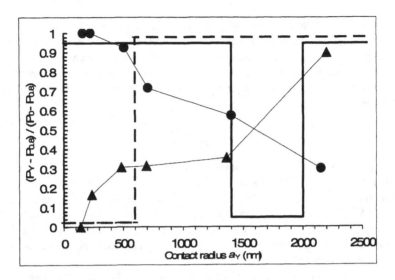

Figure 4. Normalized yield pressure ($\frac{P_Y - P_{0.8}}{P_0 - P_{0.8}}$) against contact radius at yield (a_Y).
The symbols are the same as in Fig. 3. The dotted line and solid line show
schematically the conventional prediction for the yield of the structures B1 and B2
respectively.

Only for the smallest indenter, with a contact radius at yield of about 150nm, do
the data correspond to these predictions. This indicates that for this contact radius
initial yield does occur within the depth range 0 to 300 nm. Even for the next larger
indenter, the values of *e* depart from zero and unity showing that the volume of initial
yield ranges from less than 300nm deep to more than 700nm deep. And for the sample
B1, which 300nm deep, for the larger indenters the trend of the data suggests that the
effect of the superlattice band becomes small only when a_Y reaches as much as
2000nm and will disappear only for still larger a_Y. The data for this sample strongly
suggest a minimum volume for the initial yield of at least one micron for the larger
indenters.

For the sample B2, which has the strained band at 700nm to 1000nm, the trend
suggests that the maximum effect (*e* approaching zero) is not reached with the range
of indenter sizes used here, with a_Y up to 2200 nm. It might reasonably occur for a_Y
as great as 3500 nm. Since the superlattice band is centred at a depth of 850 nm, this
is telling us that yield initiates more likely at about ¼ a_Y than at the accepted value of
½ a_Y. The broadening of the curve compared with the ideal top-hat function suggests
again that for the larger indenters yield initiates throughout a depth range from near
zero to considerably more than the depth of the layer, i.e. over a range in excess of

one micron. However, to really measure the volume, or depth range, as a function of indenter size, more mapping samples are certainly required with the strained superlattice band lying between 300nm to 700nm and deeper than 700nm.

CONCLUSIONS

These results are consistent with the results obtained by Lloyd et al [10] from cross-section TEM studies of superlattice material under the indentations. Their work was described in more detail in [4, 10]. They could only measure the plastic zone size post-yield, but they projected the plastic zone dimensions back to the initial yield point and obtained an initial yield volume of width 1.6μm and depth 1.2μm for a 5μm radius spherical indenter with a_Y approximately one micron. The dimensions, in particular the depth, are of similar magnitude to those we observed.

The results presented here demonstrate a unique experimental method in which superlattices are used to observe and map directly the real initial volume of yield for spherical indentation. The preliminary results show that when compared with conventional theory, initial yield is most likely centred at a depth of about ¼ a_Y rather than the accepted value of ½ a_Y

ACKNOWLEDGEMENTS

We are grateful to the EPSRC for financial support (under grant # EP/C518004/1).

REFERENCES

1. A. M. Minori, S. A. Syed Asif, Z. Shan, E.A.Stach, E. Cyrankowshi, T.J. Wyrobek and O. L. Warren, *Nature Materials Letters*, 5 697 (2006).
2. J.D. Kiely, K.F. Jarausch, J.E. Houston, P.E. Russell, *Journal of Materials Research*, 14 2219 (1999)
3. K.M.Y. P'ng, A.J. Bushby and D.J. Dunstan, *Mat. Res. Soc. Symp. Proc.* 778 U6.3.1 (2003)
4. N. B. Jayaweera, J. R. Downes, M. D. Frogley, M. Hopkinson, A.J. Bushby, P. Kidd, A. Kelly and D.J. Dunstan, *Proc. Roy. Soc. A.*, 459 2049 (2003)
5. N.B. Jayaweera, J.R. Downes, D.J. Dunstan, A.J. Bushby, P. Kidd and A. Kelly, *MRS Symp. Proc.* 634 B4.10 (2001)
6. M.E. Brenchley, M. Hopkinson, A. Kelly and D.J. Dunstan, *Phys. Rev. Letters* 78 3912 (1997)
7. J.S. Field and M.V. Swain *J. Mater. Res.* 8 297 (1993)
8. T.T. Zhu, A.J. Bushby and D.J. Dunstan, *Journal of Mech. Physics Solids* in press.
9. T.T. Zhu, X. Hou, A.J. Bushby and D.J. Dunstan, *Journal of Physics D* in press.
10. S. J. Lloyd, K. M.Y. P'ng, A.J. Bushby, D.J. Dunstan and W. J. Clegg, *Philosophical Magazine*, 85 2469 (2005)

Mater. Res. Soc. Symp. Proc. Vol. 1049 © 2008 Materials Research Society   1049-AA06-05

Quasi-static and Oscillatory Indentation in Linear Viscoelastic Solids

Yang-Tse Cheng[1] and Che-Min Cheng[2]
[1]General Motors R&D Center, Warren, MI, 48090
[2]Institute of Mechanics, Chinese Academy of Sciences, Beijing, 100080, China, People's Republic of

ABSTRACT

Instrumented indentation is often used in the study of small-scale mechanical behavior of "soft" matters that exhibit viscoelastic behavior. A number of techniques have been used to obtain the viscoelastic properties from quasi-static or oscillatory indentations. This paper summarizes our recent findings from modeling indentation in linear viscoelastic solids. These results may help improve methods of measuring viscoelastic properties using instrumented indentation techniques.

INTRODUCTION

Instrumented indentation [1-13] can be performed in either quasi-static or oscillatory mode for measuring mechanical properties of "soft" matters, such as polymers, composites, and biomaterials, that are often viscoelastic. In the load- or displacement-controlled quasi-static mode, the load-displacement curves are recorded. One of the widely used methods, due to Oliver and Pharr [2], is to obtain the elastic modulus from the initial unloading slope, $S = (dF/dh)_m$, at the maximum indenter displacement, h_m,

$$S = \frac{dF}{dh}\bigg|_{h=h_m} = \frac{4G}{1-v}a = \frac{2E}{\sqrt{\pi}(1-v^2)}\sqrt{A},$$ (1)

where G is the shear modulus, v is Poisson's ratio, $E = 2G(1+v)$ is Young's modulus, a is the contact radius, and $A = \pi a^2$ is the contact area. The contact radius, a, can be obtained from the contact depth, h_c, and indenter geometry. Oliver and Pharr [2] proposed an equation for h_c:

$$h_c = h_m - \xi \frac{F_m}{(dF/dh)_m},$$ (2)

where F_m is the load at h_m. The numerical value of ξ is $(2/\pi)(\pi-2) = 0.727$ and $3/4$ for a conical and paraboloid of revolution, respectively. Although Eqs. (1) and (2) were derived from solutions to elastic contact problems, they have been used for indentation in elastic-plastic solids and viscoelastic solids. One of our motivations was to evaluate whether Eqs. (1) and (2) could be used for indentation in linear viscoelastic solids and another was to improve the existing methods [14-18].

In the oscillatory mode, a sinusoidal force is typically superimposed on a quasi-static load on the indenter [1,3,4,6,9,10,13]. The indentation displacement response and the out-of phase angle between the applied harmonic force and the assumed harmonic displacement may be recorded at a given excitation frequency or multiple frequencies. Several authors [1,3,4,6,9,10,13] have proposed analysis procedures for determining the complex Young's modulus, $E^*(\omega) = E'(\omega) + iE''(\omega)$, where $E'(\omega)$ is the storage modulus and $E''(\omega)$ is the loss modulus, from oscillatory indentations using the following equations:

$$\frac{E'}{1-v^2} = \frac{\sqrt{\pi}}{2}\frac{S}{\sqrt{A}} \quad \text{and} \quad \frac{E''}{1-v^2} = \frac{\sqrt{\pi}}{2}\frac{C\omega}{\sqrt{A}}, \tag{3}$$

where v is Poisson's ratio, S is contact stiffness, C is damping coefficient, and A is contact area between the indenter and the sample. For an ideal indenter with infinite system stiffness and zero mass, the contact stiffness and damping coefficient are given by $S = |\Delta F / \Delta h|\cos\phi$ and $C\omega = |\Delta F / \Delta h|\sin\phi$, where ΔF is the amplitude of sinusoidal force with angular frequency ω, Δh is the amplitude of oscillatory displacement, and ϕ is the phase angle of the displacement response. Thus, by measuring displacement amplitude and phase angle under harmonic oscillation, the reduced storage and loss modulus, $E'/(1-v^2)$ and $E''/(1-v^2)$, can be obtained from

$$\frac{E'}{1-v^2} = \frac{\sqrt{\pi}}{2\sqrt{A}}\left|\frac{\Delta F}{\Delta h}\right|\cos\phi \quad \text{and} \quad \frac{E''}{1-v^2} = \frac{\sqrt{\pi}}{2\sqrt{A}}\left|\frac{\Delta F}{\Delta h}\right|\sin\phi. \tag{4}$$

Recently, we showed that Eq. (4) is the result of linear approximation of oscillatory indentation. By performing a nonlinear analysis, we derived the corresponding set of equations without evoking the small amplitude oscillation assumption [19].

Contact mechanics of linear viscoelastic bodies became an active area of research since the mid 1950s by the work of Lee [20], Radok [21], Lee and Radok [22], Hunter [23], Gramham [24,25], and Ting [26,27]. They have derived general equations for various contact conditions. For example, they have shown, for conical indentation in a linear viscoelastic solid with a constant Poisson's ratio, the force, $F(t)$, is given by:

$$F(t) = \frac{8\tan\theta}{\pi(1-v)}\int_0^t G(t-\tau)h(\tau)\frac{dh(\tau)}{d\tau}d\tau, \tag{5}$$

where $G(t)$ is the relaxation modulus which is related to the time-dependent Young's modulus by $E(t) = 2G(t)(1+v)$ and θ the half included angle of the indenter. When force is the independent variable, the displacement, $h(t)$ is given by:

$$h^2(t) = \frac{\pi(1-v)}{4\tan\theta}\int_0^t J_s(t-\tau)\frac{dF(\tau)}{d\tau}d\tau, \tag{6}$$

where $J_s(t)$ is the shear compliance. Eqs. (5) and (6) were derived based on the assumption of that the contact area is a monotonically increasing function of time. Under the same assumption, Ting and Gramham showed that the ratio of contact depth to indenter displacement is the same as that in the purely elastic case [24, 26],

$$\frac{h_c(t)}{h(t)} = \frac{2}{\pi}. \tag{7}$$

QUASI-STATIC INDENTATION

Conical indentation in linear viscoelastic solids: initial unloading slopes without a holding-period

We have recently shown [15] that Eqs. (5)-(7) could be used to analyze initial unloading after a loading period with a non-decreasing function $h(t)$ or $F(t)$. Specifically, the initial unloading slope is given by, using Eqs. (5)-(7),

$$\frac{dF}{dh}\bigg|_{h=h_m} = \frac{4h_c \tan\theta}{1-v} \frac{1}{J(0) - \frac{1}{v_F}\int_0^{t_m} \frac{dJ_s(\eta)}{d\eta}\bigg|_{\eta=t_m-\tau} \frac{dF(\tau)}{d\tau}d\tau}, \tag{8}$$

for load-controlled indentation with an initial unloading rate $v_F = |dF/dt|$, and

$$\frac{dF}{dh}\bigg|_{h=h_m} = \frac{4\tan\theta}{1-v}[G(0)h_c(t_m^+) - \frac{2}{\pi v_h}\int_0^{t_m} \frac{dG}{d\eta}\bigg|_{\eta=t_m-\tau} h(\tau)\frac{dh(\tau)}{d\tau}d\tau], \tag{9}$$

for displacement-controlled indentation with initial unloading displacement rate $v_h = |dh/dt|$. Eqs. (7-9) have been validated using finite element calculations for fast unloading after loading with a monotonically increasing function $h(t)$ or $F(t)$ [14,15]. Under fast unloading the second terms in Eqs. (8) and (9) are negligible, these equations become the same as Eq. (1) with $G(0) = 1/J_s(0)$ in place of G. Thus, the "instantaneous" properties, $G(0)/(1-v)$ or $E(0)/(1-v^2)$, can be obtained from either displacement- or load-controlled indentation measurements using Eqs. (7)-(9), provided that the unloading rate, v_h or v_F, is sufficiently fast. When unloading rates are sufficiently fast, the unloading slope is no longer a function of the unloading rate. Our finite element calculations suggested that "sufficiently fast" unloading could be achieved when the time duration of linear unloading was about 0.1 to 0.01 times the relaxation time of linear viscoelastic materials [15]. In practice, several indentation experiments with different unloading rates spanning several orders of magnitudes may be necessary to access whether unloading rates are fast enough. It is possible that required fast unloading is unachievable in practice. It is therefore convenient to develop techniques where an arbitrary unloading rate is sufficient to allow the determination of the instantaneous modulus. Methods of "load-hold-unload" discussed in the next section make such a measurement possible.

Conical indentation in linear viscoelastic solids: initial unloading slopes with a holding-period

We consider a load profile consisting of a loading period where the force is given by a monotonically increasing function, a "hold-at-the-peak load" period with a constant force, and an unloading period with an initial unloading rate, $v_F = |dF/dt|$. We have shown [18], using Eq. (8), that

$$\frac{(1-v)}{4a}J_s(0) = \frac{1}{dF/dh} + \frac{dh/dt|_{t=t_m^-}}{v_F}. \tag{10}$$

Thus, the instantaneous properties, $G(0)/(1-v)=1/[J_s(0)(1-v)]$, can be obtained from the measurement of initial unloading slope, dF/dh, the velocity of the indenter immediately before unloading, $(dh/dt)_{t=t_m^-}$, the rate of unloading, v_F, and the contact radius a. Eq. (10) was first suggested by Ngan and co-workers [12]. Eq. (10) shows that, under load-control, the "hold-at-the-peak-load" method provides a convenient means to determine the instantaneous modulus. In particular, when the holding period is sufficiently long, the ratio of $(dh/dt)_{t=t_m^-}$ over v_F becomes negligibly small as a result of creep. The instantaneous modulus can then be obtained directly from the unloading slope dF/dh and the contact radius or depth.

We have also proposed a "hold-at-the-maximum-displacement" method for indentation measurements when displacement is the independent variable [18]. We considered a displacement profile where the displacement is given by a monotonically increasing function, a hold period at a constant displacement, and an unloading period with an initial unloading rate, $v_h = |dh/dt|$. Using Eq. (9), we have shown [18] that

$$\frac{4G(0)}{1-v}a = \frac{dF}{dh}\Big|_{h=h_m} + \frac{dF/dt\Big|_{t_m^-}}{v_h}. \tag{11}$$

Eq. (11) shows that $G(0)/(1-v)$ can be obtained by measuring the initial unloading slope, dF/dh, the rate of force relaxation immediately before unloading, $(dF/dt)_{t=t_m^-}$, the rate of unloading, v_h, and the contact radius a. When the holding period is sufficiently long, the ratio of $(dF/dt)_{t=t_m^-}$ over v_h becomes negligibly small as a result of relaxation. The instantaneous modulus can be obtained directly from the unloading slope dF/dh and the contact radius or depth.

We found, using finite element calculations [15, 16,18], that the Oliver-Pharr equation for contact depth (Eq. (2)) often produces significant errors, whereas Eq. (8) is indeed a good approximation for the contact depth up to initial unloading for conical indentation. The reason that Eq. (2) is not applicable to indentation in viscoelastic solid becomes clear if we examine the two relationships used in deriving Eq. (2): $h_c/h = 2/\pi$ and $h = 2F/(dF/dh)$ for conical indentations. The first equation is identical to Eq. (7). However, the second equation, which comes from $F = Ch^2$ where C is a time-independent parameter, is, in general, not true for conical indentation in linear viscoelastic solids as can be seen from either Eq. (5) or (6) for the displacement- and load-controlled indentation, respectively. Thus, $E(0)/(1-v^2)=2G(0)/(1-v)$ can be obtained using Eqs. (10) or (11) together with Eq. (8) in "load-hold-unload" measurements.

OSCILLATORY INDENTATION

In oscillatory indentation measurements, a harmonic force is superimposed on a quasi-static force, i.e.,

$$F(t) = F_m f(t) + \Delta F \sin(\omega t), \qquad (12)$$

where $f(t)$ is a monotonically non-decreasing function of time $|f(t)| \leq 1$. Inserting Eq. (12) into (6) and using the definition of the storage and loss shear compliances, $J'(\omega) = \omega \int_0^\infty J_s(s) \sin(\omega s) ds$ and $J''(\omega) = -\omega \int_0^\infty J_s(s) \cos(\omega \cdot s) ds$, we obtain [19]:

$$h^2(t) = \frac{\pi(1-\nu)}{4 \tan \theta} \left\{ F_m \int_{-\infty}^t J_s(t-\tau) \frac{df(\tau)}{d\tau} d\tau + J'(\omega) \Delta F \sin(\omega \cdot t) - J''(\omega) \Delta F \cos(\omega \cdot t) \right\}. \qquad (13)$$

Eq. (13) shows that $h^2(t)$ can be expressed as,

$$h^2(t) = B(t) + \Delta_2 h \sin(\omega t - \phi), \qquad (14)$$

where $\Delta_2 h$ is the "square" amplitude of harmonic displacement and ϕ is the phase shift. Eq. (14) shows that the square of displacement, $h^2(t)$, is a sinusoidal function of time. This is different from the usual assumption that the displacement, $h(t)$, is a sinusoidal function of time based on which Eqs. (3-4) were derived. Comparing Eq. (14) with Eq. (13) and using the relationship between the complex modulus, E^*, and the complex shear compliance, J^*, i.e.,

$E' = 2(1+\nu) \dfrac{J'}{J'^2 + J''^2}$ and $E'' = 2(1+\nu) \dfrac{J''}{J'^2 + J''^2}$, we obtained [19],

$$B(t) = \frac{\pi(1-\nu)}{4 \tan \theta} F_m \int_{-\infty}^t J_s(t-\tau) \frac{df(\tau)}{d\tau} d\tau, \qquad (15)$$

$$E' = \frac{\pi(1-\nu^2)}{2 \tan \theta} \frac{\Delta F}{\Delta_2 h} \cos\phi \quad \text{and} \quad E'' = \frac{\pi(1-\nu^2)}{2 \tan \theta} \frac{\Delta F}{\Delta_2 h} \sin\phi. \qquad (16)$$

Eq. (16) shows that the storage and loss modulus can be obtained by measuring the square amplitude of displacement $\Delta_2 h$ and phase shift ϕ. The measurement of the contact area or the absolute position of the indenter is unnecessary, thus removing the difficulties associated with contact area measurement and thermal drift for both large and small amplitude oscillatory indentations.

CONCLUSIONS

We have provided an overview of our recent studies of indentation in linear viscoelastic solids. These studies established basic equations for several methods of obtaining viscoelastic properties using quasi-static and oscillatory indentations. Specifically, we showed that the instantaneous modulus, $E(0)/(1-\nu^2) = 2G(0)/(1-\nu)$, can be obtained using either the method of fast unloading or the method of load-hold-unload for both load- and displacement-controlled quasi-static measurements. We also derived equations for obtaining the storage and loss modulus from oscillatory indentation measurements without using the usual assumption of small amplitude oscillations. Although we focused our discussions on conical indentations in this overview, corresponding equations have been derived for quasi-static and oscillatory indentations in linear viscoelastic solids using spherical indenters [16,19]. We hope these results will help improve the current practices of using indentation to determine viscoelastic properties of "soft" materials, including polymers, composites, and biomaterials.

ACKNOWLEDGMENTS

We would like to thank Dr. Wangyang Ni for his contributions to the work summarized in this paper. We would also like to thank Mike Lukitsch, Yue Qi, Tom Perry, and Mark W. Verbrugge for valuable discussions. C.-M. Cheng would like to acknowledge partial support from NSF of China, Project No.10372101.

REFERENCES

1. J. B. Pethica and W. C. Oliver, Phys. Scr. T19, 61 (1987).
2. W. C. Oliver and G. M. Pharr, J. Mat. Res. 7, 1564 (1992).
3. J. L. Loubet, B. N. Lucas, and W. C. Oliver, in International Workshop on Instrumental Indentation, San Diego, CA, April 1995, D.T. Smith (ed.), 31 (1995).
4. S. A. Syed, K. J. Wahl, and R. J. Colton, Rev. Sci. Instrum. 70, 2408 (1999).
5. S. Shimizu, T. Yanagimoto, M. Sakai, J. Mat. Res. 14, 4075 (1999).
6. N. A. Burnham, S. P. Baker, and H. M. Pollock, J. Mat. Res. 15, 2006 (2000).
7. L. Cheng et al., J. Poly. Sci.: Part B: Poly. Phys. 38, 10 (2001).
8. M. L. Oyen and R. F. Cook, J. Mat. Res. 18, 139 (2003).
9. M. R. VanLandingham, J. Res. Nat. Inst. Stand. Tech. 108, 249 (2003).
10. A.C. Fischer-Cripps, Nanoindentation, 2nd edition (Springer-Verlag, New York, 2004).
11. G. Huang, B. Wang and H. Lu, Mech. Time-Dependent Mater. 8, 345 (2004).
12. A.H.W. Ngan, H.T. Wang, B. Tang, K.Y. Sze, Int. J. Solids and Struct. 42, 1831 (2005).
13. G. M. Odegard, T. S. Gates and H. M. Herring, Exp. Mech. 45, 130 (2005).
14. Y.-T. Cheng and C.-M. Cheng, Mat. Sci. Eng. R44, 91 (2004).
15. Y.-T. Cheng and C.-M. Cheng, J. Mat. Res. 20, 1046 (2005).
16. Y.-T. Cheng and C.-M. Cheng, Materials Science and Engineering A 409, 93 (2005).
17. Y.-T. Cheng and C.-M. Cheng, Appl. Phys. Lett. 87, 111914 (2005).
18. Y.-T. Cheng, W. Ni, and C.-M. Cheng, J. Mat. Res. 20, 3061 (2005).
19. Y.-T. Cheng, W. Ni, and C.-M. Cheng, Phys. Rev. Lett. 97, 075506 (2006).
20. E. H. Lee, Quarterly Appl. Math. 13, 183 (1955).
21. J. R. M. Radok, Quarterly Appl. Math. 15, 198 (1957).
22. E. H. Lee, J. R. M. Radok, J. Appl. Mech. 27, 438 (1960).
23. S. C. Hunter, J. Mech. Phys. Solids 8, 219 (1960).
24. G. A. C. Graham, Int. J. Eng. Sci. 3, 27 (1965).
25. G. A. C. Graham, Int. J. Eng. Sci. 5, 495 (1967).
26. T. C. T. Ting, J. Appl. Mech. 33, 845 (1966).
27. T. C. T. Ting, J. Appl. Mech. 35, 248 (1968).

Mater. Res. Soc. Symp. Proc. Vol. 1049 © 2008 Materials Research Society   1049-AA06-06

Indentation of Nonlinearly Viscoelastic Solids

Michelle L Oyen
Engineering Department, Cambridge University, Trumpington Street, Cambridge, United Kingdom

ABSTRACT

Much recent attention has been focused on the indentation of linearly viscoelastic solids, and analysis techniques have been developed for polymeric material characterization. However, there has been relatively little progress made in the development of analytical approaches for indentation of nonlinearly viscoelastic materials. Soft biological tissues tend to exhibit responses which are nonlinearly viscoelastic and are frequently modeled using a decomposition of the relaxation or creep function into a product of two functions, one time-dependent and the other stress- or strain-level dependent. Consideration here is for soft biological tissue-like responses, exhibiting approximately quadratic stress-strain behavior, which can be alternatively cast as linear dependence of elastic modulus on strain level. An analytical approach is considered in the context of indentation problems with flat punch, spherical and conical indenter shapes. Hereditary integral expressions are developed and solved for typical indentation experimental conditions including indentation creep, load-relaxation and monotonic constant load- or displacement-rate tests. Primary emphasis is on the deconvolution of material and geometrical nonlinearities during an indentation experiment. The simple analytical expressions that result from this analysis can be implemented for indentation characterization of soft biological tissues without the need for computationally- intensive inverse finite element approaches.

INTRODUCTION

Depth-sensing (instrumented) indentation testing, in which load and displacement are monitored during contact of a probe with a material surface, is a popular technique for measurement of local mechanical properties of a wide variety of engineering materials. This has been due in part to the wide-spread availability of commercial instruments for small-scale contact testing ("Nanoindentation") along with the development of routine analytic technique for elastic-plastic [1,2], viscoelastic [3,4], poroelastic [5] and viscous-elastic-plastic [6,7] mechanical property deconvolution.

Because of the capability for localized testing, nanoindentation testing is particularly well-suited to the mechanical analysis of biological materials, in which the mechanical properties can vary substantially from point to point [8]. These property variations result from local variations in tissue composition and microstructure, and the variations are frequently associated with the length-scales of extracellular matrix components (i.e. length scales of tens of nanometers to micrometers) and of cell activity (i.e. length scales of micrometers to tens of micrometers).

In common practice, the extensive load-displacement (P–h) responses are nonlinear under conditions of spherical or conical-pyramidal indentation even when the material is linearly elastic ($\sigma = E\varepsilon$). This nonlinearity is geometrical: as the indenter is pushed further into the material, the contact area increases giving rise to responses of the form $P \propto h^m$, $m \neq 1$. Indentation analysis for linearly viscoelastic responses can also be considered in the context of geometrical nonlinearity [9].

A complication arises in the application of nanoindentation techniques for examination of soft biological materials, in which the material responses are nonlinearly viscoelastic [10]. In this circumstance, even under homogeneous loading conditions and time-independence, the load-displacement responses are nonlinear. Therefore, under indentation conditions (with a spherical or conical-pyramidal indenter) the possibility exists for simultaneous nonlinear effects due to material and geometry. In the current work, this dual nonlinearity condition is explored analytically.

THEORY

Linear Viscoelasticity

A Boltzmann hereditary integral formulation can be used to solve for the time-dependent response of a material to any arbitrary mechanical loading history. Here we use a common simplifying assumption that volumetric responses are time-independent while deviatoric responses are time-dependent [11]. In strain-control, we can then write the time-dependent deviatoric stress $s_{ij}(t)$ as a function of the shear relaxation function $G(t)$ and applied strain history $e_{ij}(t)$ by the integral formulation:

$$s_{ij}(t) = 2 \int_0^t G(t-u) \frac{de_{ij}(u)}{du} du \tag{1}$$

A linearly elastic material with a Maxwell (Fig. 1a) response has a relaxation function,

$$G(t) = G \exp\left(\frac{-t}{\tau_M}\right), \quad \tau_M = \frac{\eta}{G} \tag{2}$$

and for a standard linear solid (Fig. 1b) the relaxation function is,

$$G(t) = \frac{G_1}{G_1 + G_2}\left[G_2 + G_1 \exp\left(\frac{-t}{\tau_{RS}}\right)\right], \quad \tau_{RS} = \frac{\eta}{G_1 + G_2} \tag{3}$$

Integration of Eqn. 1 can take place over the different relaxation functions (Eqns 2 or 3) to obtain the appropriate closed-form expression for the stress-strain-time response.

Maxwell

(a)

Standard Linear Solid

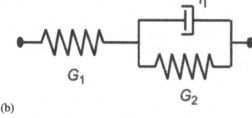

(b)

Figure 1: Spring and Dashpot models for simple linearly viscoelastic material models (a) Maxwell fluid and (b) standard linear solid.

Geometrical Nonlinearity

Under indentation loading conditions, the elastic problem simplifies to a closed-form expression relating axial load (P) and axial displacement (h), as demonstrated here for spherical indentation [11] with a sphere radius of R:

$$P = \frac{4\sqrt{R}}{3}\frac{E}{1-v^2}h^{3/2} = \frac{8\sqrt{R}}{3}(2G)h^{3/2} \qquad (4)$$

where in the second instance incompressibility ($v = 0.5$) has been assumed in substituting the shear modulus (G) for the elastic modulus (E). For conical indentation the P-h expression is written:

$$P = \frac{\pi\tan\psi}{2\gamma^2}\frac{E}{(1-v^2)}h^2 = \frac{\pi\tan\psi}{\gamma^2}(2G)h^2 \qquad (5)$$

for a conical indenter with included angle ψ and where γ is the ratio of total depth to contact depth (normally $\gamma = \pi/2$). In both equations 4 and 5, there is a clear geometrical nonlinearity which is defined by the fact that the relationship between load and displacement is for a power not equal to one. This case has been considered for a viscoelastic material, in which for spherical indentation [9, 11, 12] (combining the approach of Eqn. 1 with Eqn. 4) so long as the contact area is non-decreasing:

$$P = \frac{8\sqrt{R}}{3} 2Gh^{3/2} \rightarrow P(t) = \frac{8\sqrt{R}}{3} \int_0^t G(t-u) \frac{dh^{3/2}(u)}{du} du \tag{6}$$

which can be integrated over a linearly viscoelastic relaxation function (Eqns 2,3) and an applied arbitrary displacement history $h(t)$ to find the load-displacement-time (P-h-t) response in closed form if a closed-form expression exists. Because of the $dh^{3/2}(u)/du$, obvious experimental conditions such as constant displacement rate [$h(t) = kt$] do not give rise to closed-form expressions [12] for the typical exponential relaxation functions (Eqns 2,3).

Material Nonlinearity

For nonlinearly viscoelastic materials, the relaxation function depends on both strain level and on time. A common assumption is that the strain- and time-dependent components are separable, such that the relaxation function can be written [13,14]:

$$G(\varepsilon, t) = f_1(\varepsilon) f_2(t) \tag{7}$$

in which $f_2(t)$ is a normalized ("reduced") relaxation function with $f_2(0) = 1$ and $0 \le f_2(\infty) < 1$. The strain-dependent term is the first derivative of the stress-strain response,

$$f_1(\varepsilon) = \frac{d\sigma(\varepsilon)}{d\varepsilon} \tag{8}$$

such that for a linearly viscoelastic case (based on a linearly elastic $\sigma = E\varepsilon$ relationship) $f_1 = a$ constant. For the Maxwell model, a reduced relaxation function of $f_2 = \exp(-t/\tau)$ and a linearly elastic shear stress-shear strain relationship ($f_1 = G$) recovers the original relaxation function (Eqn. 2). For higher order power-law stress-strain relationships, such as $\sigma = D\varepsilon^2$ or $\sigma = F\varepsilon^3$, the strain-dependent functions are linear and quadratic in strain, respectively. The hereditary integral formulation for material nonlinearity is written,

$$\sigma(t) = \int_0^t f_2(t-u) f_1(\varepsilon) \frac{d\varepsilon}{du} du = \int_0^t f_2(t-u) \frac{d\sigma(\varepsilon)}{d\varepsilon} \frac{d\varepsilon}{du} du \tag{9}$$

and solved by integration for an arbitrary strain history $\varepsilon(t)$.

Nonlinearity Comparisons

An interesting coincidence occurs in the comparison of a geometrically nonlinear case of conical indentation expression, where $P \propto h^2$, and the materially nonlinear case of $\sigma = D\varepsilon^2$. The solution for load- or stress-time (P-t or σ-t) histories involves the same functional form within the hereditary integral, as demonstrated in Figure 2.

For constant ramp rate, k ($d\varepsilon/dt$ or dh/dt):

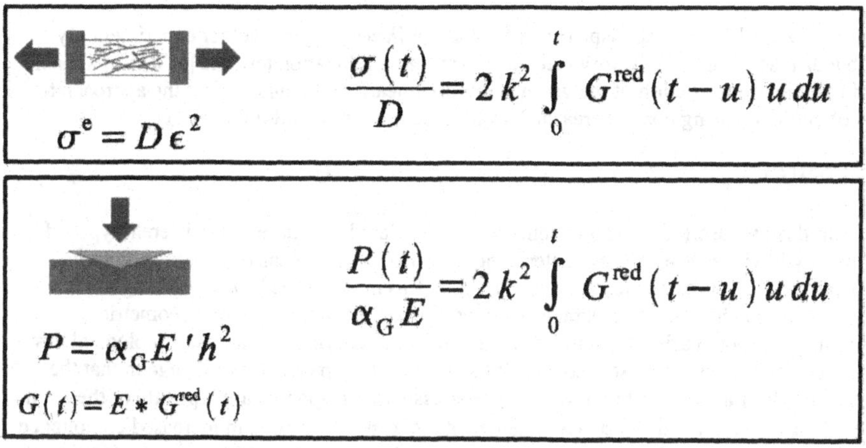

$$\frac{\sigma(t)}{D} = 2k^2 \int_0^t G^{red}(t-u)\,u\,du$$

$$\sigma^e = D\epsilon^2$$

$$\frac{P(t)}{\alpha_G E} = 2k^2 \int_0^t G^{red}(t-u)\,u\,du$$

$$P = \alpha_G E' h^2$$

$$G(t) = E * G^{red}(t)$$

Figure 2: Comparison of quadratic nonlinearity cases for material nonlinearity (top) and geometrical nonlinearity (bottom).

Combined Nonlinearity

For indentation of a nonlinearly viscoelastic material, the approaches of hereditary integrals for geometrical (Eqn. 1) and material (Eqn. 9) nonlinearity will be combined. To write the material nonlinearity expression in terms of indentation coordinates, for spherical indentation we use the expression approximating strain in terms of contact radius (a) and sphere radius (R) or indentation displacement (h) and indenter radius as [11]:

$$\varepsilon = 0.2a/R = 0.2(h/R)^{1/2} \Rightarrow \varepsilon = f_3(h^{1/2}) \tag{10}$$

where here f_3 is the constant 0.2. Because of the geometric similarity in a conical indenter, the indentation strain is a constant function of the indenter included angle (ψ) and the sensitivity to material nonlinearity may be diminished. We thus focus on spherical indentation for this application.

The relaxation function in strain-time $G(t,\varepsilon)$ (Eqn. 7) becomes a function $G(t,h)$ of time and the extensive coordinate h via Eqn. 10 such that:

$$G(t,h) = f_2(t)f_3(h^{1/2}) \tag{11}$$

The hereditary integral for load becomes

$$P(t) = \frac{8R^{1/2}}{3} \int_0^t f_2(t-u) f_3\left(h^{1/2}(u)\right) \frac{dh^{3/2}(u)}{du} \, du \qquad (12)$$

this can be solved for a ramp displacement condition $[h(u) = ku]$ or another experimentally sensible displacement-time history (such as a ramp-hold relaxation test, [12]). It is reasonable to remain within the restriction of the geometrical nonlinearity case and restrict the approach to cases of non-decreasing contact area such as load-hold but not load-unload tests.

DISCUSSION

In this manuscript, two approaches were considered for addressing different types of nonlinearity in viscoelastic solids: material nonlinearity, which is intrinsic, and geometric nonlinearity, which is extrinsic. A novel approach was then outlined for combining the two cases into a single hereditary integral expression for combined material and geometric nonlinearity, as would arise when indenting a nonlinear material such as a soft biological tissue. In practice, the nonlinearly viscoelastic soft tissue presents a particular challenge in that the response is stiffer at larger strain levels, as the stress-strain responses are typically of the form $\sigma = D\varepsilon^m$, $m > 1$. However the strong time-dependent response results in increased compliance with increased experimental time such that this softening competes with increased effective stiffness at larger strain levels. As such, a substantial battery of experiments conducted over different strain-time or displacement-time histories is required to identify unique nonlinearly viscoelastic material properties from indentation experiments.

REFERENCES

1. Field JS and Swain MV, *J Mater Res* **8** (1993) 297-306.
2. Oliver WC and Pharr GM, *J Mater Res* **7** (1992) 1564-83.
3. Oyen ML, *Philosophical Magazine* 2006, **86** 5625 - 5641.
4. Oyen ML, *Journal of Materials Research*, **20** (2005) 2094-2100.
5. Oyen ML, Bembey AK, Bushby AJ, in Mechanics of Biological and Bio-Inspired Materials, edited by C. Viney, K. Katti, C. Hellmich, U. Wegst (Mater. Res. Soc. Symp. Proc. 975E, Warrendale, PA, 2007), 0975-DD07-05.
6. Oyen ML and Cook RF: *Journal of Materials Research*, **18** (2003) 139-50.
7. Cook RF and Oyen ML, *International Journal of Materials Research*, **98** (2007) 370-8.
8. Oyen ML and Ko C-C, *J. Mater. Sci. Mater. Med.*, **18** (2007) 623-8.
9. Lee EH and Radok JRM, *J. Applied Mech.* **27** (1960) 438-44.
10. Oyen ML, Cook RF, Stylianopoulos T, Barocas VH, Calvin SE, and Landers DL, *Journal of Materials Research*, **20** (2005) 2902-9.
11. Johnson KL, Contact Mechanics. (Cambridge University Press, Cambridge UK, 1985).
12. Mattice JM, Lau AG, Oyen ML, Kent RW, *J. Mater. Res.* **21** (2006) 2003-10.
13. Findley WN, Lai J, and Onaran K, *Creep and Relaxation of Nonlinear Viscoelastic Materials*, (Dover, New York, 1989).
14. Fung YC, *Biomechanics: Mechanical Properties of Living Tissues*, 2nd edition, (Springer-Verlag, New York, 1993).

Mater. Res. Soc. Symp. Proc. Vol. 1049 © 2008 Materials Research Society   1049-AA06-08

Determination of Residual Stress and Yield Stress Simultaneously by Indentation Method with Dual Indenters

Baoxing Xu, Xinmei Wang, and Zhufeng Yue
School of Mechanics, Civil Engineering and Architecture, Northwestern Polytechnical University, Xi'an, 710072, China, People's Republic of

ABSTRACT

Obtaining residual stress from indentation test requires calculating the contact area between the indenter and the indented material, while yield stress of indented material is to be known in advance. In this work, the dimensional analysis of indentation loading curve was first analyzed, and then a reverse numerical procedure was explored to determine the residual stress and yield stress of materials simultaneously from indentation test. Besides, the calculation of contact area can be also avoided.

INTRODUCTION

The measurement of residual stress in the surface of materials with indentation test has attracted intensive interests since Sines and Carlson suggested that the residual stress could be measured by utilizing effects of residual stress on the hardness [1]. Recently, several methods have been explored to extract residual stress from indentation load-depth curves [2-6].

Inspired from results of Tsui et al. that residual stress affected the measurement of hardness and elastic modulus using a Berkovich indenter [7], Suresh and Giannakopoulos suggested a method of determination of equi-biaxial residual stress utilizing difference in contact area of stressed and unstressed materials at the same depth with Berkovich indenters [2]. But in fact, the influence of residual stress on contact area is relatively small for the Berkovich indenter, especially when the residual stress is close to yield stress and the pile-up is obvious in materials. Later, Taljat and Pharr suggested that much larger effects from residual stress can be obtained by using blunt, spherical indenters [3]. They performed spherical indentation test on a polished disk of commercial aluminum alloy, which was pre-applied compressive or tensile biaxial stress, and found that indentation load - depth curves shifted to a larger indentation depth when the applied biaxial stress was tensile stress, while compressive stress had opposite effect. Based on these experimental results, Swadener et al put forward a method of measuring residual stresses using spherical indenters [4]. However, this method has the disadvantage that contact area has to be calculated, which is difficult due to the influence of pile- up around indention. Besides, the yield stress of materials needs to be known in advance. Later, in order to avoid the calculation of contact area, Xu et al. give a similar method by using two different sizes of flat cylindrical indenters. But the yield stress still needs to be known in advance [8].

For the indentation test, the unique advantages of dual indenters have been reviewed recently by Chen et al. [9]. In the present study, we show the possibility that residual stress and yield stress can be determined simultaneously with dual indenters from load – depth curves of indentation test with the aid of a set of finite element calculations.

THEORY AND NUMERICAL MODEL

Dimensional analysis

Cheng and Cheng have outlined an overall framework of applying dimensional function to indentation test in order to obtain materials properties [10-11]. Here, we give a simple dimensional analysis of indentation test when a uniform residual stress σ^R exists in indented materials. The geometry of indenter is sharp with a half angle θ, and its deformation is rigid. The friction coefficient at the contact surface between the indenter and the indented material is assumed to be zero. The indented material is assumed to be a homogeneous elastic-plastic bulk material under uniform residual stress σ^R applied, and its stress- strain relationship can be represented by the power hardening law:

$$\sigma = \begin{cases} E\varepsilon & \sigma < \sigma_o \\ K\varepsilon^n & \sigma \geq \sigma_o \end{cases} \tag{1}$$

where E is elastic modulus and σ_o is yield stress; K is the strength coefficient and n is the work hardening exponent.

Applying the Π dimensionless function, the indentation load F can be written:

$$F = Eh^2\Pi(\sigma^R / E, \sigma_o / E, n, v, \theta) \tag{2}$$

And thus

$$C = F / h^2 = E\Pi(\sigma^R / E, \sigma_o / E, n, v, \theta) \tag{3}$$

From Eq.(3), the parameter C is dependent on yield stress σ_o, residual stress σ^R and half angle of indenter θ if other parameters are given. In our prior indentation creep analysis, we put forward a method that two parameters can be obtained by changing parameters that are related with the geometry of indenter [12-13]. Here, we show that this method can be used to determine residual stress and yield stress.

FEM model

An axisymmetric two-dimensional finite element model is constructed here to save computational time. The finite element meshes are four-noded isoparametric quadrilateral elements, and they are gradually refined from the outer boundary towards the indented zone. In addition, the meshes are further refined around the edge of the contact zone for the large deformation and steel stress gradients at the edge of the contact region to accurately simulate deformation process in the region of contact. The indenter has a conical tip with a half-included angle θ. Two sizes of $\theta = 60^o$ and $\theta = 70.3^o$ will be chosen.

Axisymmetric boundary conditions are applied on the nodes along the centerline of the indenter and indented material. Roller boundary conditions are applied on nodes at the bottom of indented material. A radial stress was pre-applied to simulate a uniform residual stress on the outer cylindrical surface of indented material. For symmetry, the indented material is therefore contained in uniform residual stress before indentation test. The finite element software ABAQUS is used to performed all finite element calculations [14].

The deformation of indented material obeys isotropic power hardening law shown in Eq.(1). The basic premises here is that elastic modulus E and hardening exponent n are 72.36GPa and 0.05 respectively, allowing us to investigate the effect of both yield stress and residual stress in a straightforward and efficient way. The feasibility of this assumption has been verified on reverse problems by Yan et al. [15].

FEM RESULTS

Indentation loading curve and influence of residual stress

Fig.1 shows the influence of residual stress σ^R on indentation loading curves. The residual stress σ^R varies from -200MPa to 200MPa (minus sign denotes compression). It can be seen that indentation loading curves is in good agreement with dimensional analysis Eq.(3) with a constant slope that is the ratio of indentation load to the square of indentation depth. Indentations load increases with increase of compressive residual stress and with decrease of tensile residual stress when other conditions are same.

The effect of residual stress on indentation loading curves can be explained from the view of shear plasticity. Since the indentation stress is compressive and perpendicular to the applied surface, the existence of tensile residual stress in the material will increase the magnitude of shear stress under indenter. As a result, the plastic deformation in the material is enhanced resulting in a greater indentation load. On the contrary, the compressive residual stress has an opposite effect. A simplified theoretical model for explanation has been given in our pervious research [8].

To further show the influence of residual stress σ^R on analysis constant C, Fig.2 gives C Vs σ^R curves with $\sigma_o = 300MPa$ and $\theta = 70.3^o$. From Fig.2, two characteristics can be seen obviously. First, there is an increase for C with increase of compressive residual stress and a decrease with increase of tensile residual stress. Second, compressive residual stress has a larger effect on C over tensile residual stress. That is to say, there is a greater slope of $C - \sigma^R$ curves at the stage of compressive residual stress, which is in agreement with the influence of residual stress on the indentation contact hardness [1].

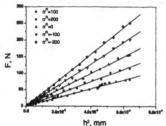

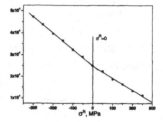

Figure 1. Indentation load – depth curve with residual stress varied from -200MPa to 200MPa, FE calculation parameters are $\sigma_o = 300MPa$ and $\theta = 70.3^o$.

Figure 2. Influence of residual stress σ^R on analysis constant C with parameters $\sigma_o = 300MPa$ and $\theta = 70.3^o$.

Determination of residual stress and yield stress simultaneously from indentation test

Eq.(3) shows that it is possible to determine residual stress and yield stress by using analysis constant C as a transition from indentation loading curves. In this section, we will put forward a method to show this possibility by combining FE calculations and indentation test. Here, we assume that there exists a compressive residual stress in material in order to simplify FE calculation process. First, a set of analysis constant C Vs compressive residual stress σ^R curves can be obtained by a set of assumed yield stress values for $\theta = 70.3°$, as shown in Fig.3. Second, we now draw a horizontal line with an analysis constant value of C=53193.1(corresponding to the residual stress σ^R and yield stress σ_o at -180MPa and360MPa, respectively), which is obtained from indentation test performed with the same half angle of indenter $\theta = 70.3°$ with FE calculations. Five intersections at different positions are obtained in C-σ^R curves, and we plot the corresponding pairs of σ_o and σ^R data in Fig. 4. Third, the same procedure for $\theta = 60°$ is carried out to include the resulting σ_o-σ^R data into Fig.4. Fourth, since we consider only one material characterized by only one σ^R and one σ_o value, the intersection point obtained from two half angles of indenters ($\theta = 70.3°$ and $\theta = 60°$) intersect will yield the results which we are looking for. Finally, to make accurate results, it is necessary to start the calculation by considering a narrower range of σ_o and σ^R values and use the results to refine the calculation till we get a convergence values. Here, three iterative FE calculations are given as in Table 1.

Synthesizing the above process, it can be prepared a flow diagram to further clarify the evolution of this method as Fig.5.

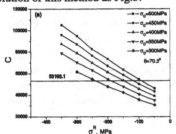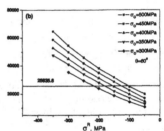

Figure 3. The map of residual stress σ^R, yield stress σ_o and analysis constant C for two different half angles of sharp indenters (a)$\theta = 70.3°$; (b)$\theta = 60°$

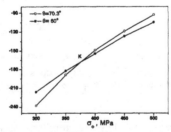

Figure 4. The curves of two sets of intersections in Fig. 3, and the intersection in this figure will result in the values of σ_o and σ^R

Table 1: Numerical results of three iterative FE calculations

	The first calculation results	The third calculation results
Indenter1: $\theta = 70.3°$ Indenter2: $\theta = 60°$	$\sigma_o = 376.6MPa$ $\sigma^R = -167.5MPa$	$\sigma_o = 368.1MPa$ $\sigma^R = -172.3MPa$
Theoretical values	$\sigma_o = 360MPa$ $\sigma^R = -180MPa$	

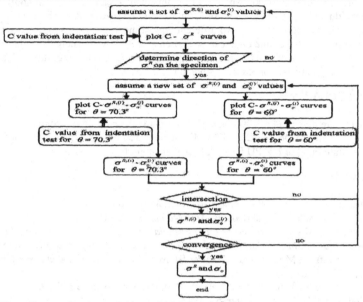

Figure 5. Flow diagram for determining residual stress σ^R and yield stress σ_o. The i represents the iterative number of FE calculations (i=1,2,3,...).

SUMMARY AND CONCLUSIONS

In the present study, we show a procedure where residual stress and yield stress can be determined simultaneously from indentation test with dual sharp indenters. This procedure requires only two tests which differ in half angle of sharp indenters but more efforts are associated with FE calculations. (1) a serial finite element calculation is performed to construct a map of residual stress, yield stress and dimensional analysis constant with coarsely pre - evaluated residual stress and yield stress. (2) Determining the sign of residual stress, i.e. compressive or tensile residual stress, by combining the indentation experimental result and the map. (3) Further finite element calculations are carried out by limiting the range of residual stress to tensile or compressive state, and a new map of residual stress, yield stress and analysis constant is built.(4) The relationship of yield stress and residual stress can be obtained by using cross-points of the indentation experimental results and the new map.(5)Yield stress and residual stress can be determined simultaneously by using the cross-points of two different relationships of yield stress and residual stress, which is built with two different half angles of indenters. It should be emphasized that Young's modulus and hardening properties are required for this method. This assumption is feasible for many alloys materials, where the alloy element can hardly change modulus but it can significantly change yield stress and residual stress. This method can also be applied to coatings (substrate effect and size effect can be neglected) because the applied residual stress is uniform here before indentation test and coating's modulus remain close to bulks counterpart, while plastic properties vary.

Provided yield stress is known in advance, this method can be greatly simplified to determine residual stress with only a set of assumed residual stresses in the FE calculations. This method can also be carried out with two different type of indenters, that is to say, one indenter geometry is a sharp indenter, the other can be chosen as a Berkovich triangular indenter. However, the uniqueness of this method may depend on angle difference of dual indenter, as evidenced in ref. [9] and that needs to be studied in the future.

ACKNOWLEDGMENTS

Thanks are given to Prof. Xi Chen of Columbia University for his constructive suggestions, and also to the NSFC (10472094), the Research Fund for the DPHE (N6CJ0001) and the Doctorate Fund of NWPU for their financial supports.

REFERENCES

1. G. Sines, and R. Carlson. *Bull. Amer. Soc. Test. Mater.* 180, 35 (1952).
2. S. Suresh, and A. E. Giannakopoulos. *Acta.Mater.*46, 5755 (1998).
3. B.Taljat and G.M.Pharr. *Measurement of Residual Stresses by Load and Depth Sensing Spherical Indentation, in Thin Films: stress and mechanical Properties VIII,* edited by R. Vinic, O. Kraft, N. Moody, and E. Shaffer III, Materials Research Society, Warrendale, PA, 2000, pp.519-524.
4. J. G. Swadener, B. Taljat, and G.M. Pharr. *J. Mater. Res.* 16, 2091 (2001).
5. Z.H. Xu, and X.D. Li. *Acta Mater.* 53, 1913 (2005).
6. X. Chen, J. Yan, and A. M. Karlsson. *Mater. Sci. & Eng. A* 416, 139 (2006).
7. T.Y. Tsui, W.C. Oliver, and G.M. Pharr. *J. Mater. Res.* 11, 752 (1996).
8. B.X. Xu, B.Zhao, and Z.F. Yue. *J. Mater. Eng. & Performance.* 15, 299 (2006).
9. X. Chen, N. Ogasawara, M.H. Zhao, and N. Chiba. *J. Mech. &Phys. Solids.* 55, 1618 (2007)
10. Y.T. Cheng, and C.M. Cheng. *Mater. Sci. & Eng. R* 44, 91 (2004).
11. Y.T. Cheng, and C.M. Cheng. *Appl. Phys. Lett.* 73, 614 (1998).
12. B.X. Xu, X.M. Wang, B. Zhao and Z.F. Yue. *Mater. Sci. & Eng. A* (2007) (in Press).
13. B.X. Xu, Z.F Yue, and G. Eggeler. *Acta Mater.* 55, 6275 (2007).
14. Hibbit, Karlsson and Sorensen. *ABAQUS Theory Manual*, Version 6.3, 2001.
15. J. Yan, A.M. Karlsson, and X. Chen. *Int. J. Solids & Struct.* 44, 3720 (2007).

Mater. Res. Soc. Symp. Proc. Vol. 1049 © 2008 Materials Research Society   1049-AA06-09

Instrumented Indentation Contact with Sharp Probes of Varying Acuity

Dylan J. Morris

Materials Science and Engineering Laboratory, National Institute of Standards and Technology, 100 Bureau Drive, MS 8520, Gaithersburg, MD, 20899

ABSTRACT

While elastic and plastic material property extraction from instrumented indentation tests has been well-studied, similarly-based fracture property measurement remains difficult. Furthermore, estimation of the fracture toughness requires measurement of the crack lengths from a micrograph, which makes nano-scale indentation toughness measurement expensive and difficult. Initiation and propagation of cracks on the nano-scale requires a more acute indenter than a Berkovich or sphere, such as the cube-corner pyramid. Experiments described here were performed on a range of elastic, plastic and brittle materials with diamond indenters of acuity varying between the Berkovich and the cube-corner. These experiments reveal some of what is changed and what remains the same, when the acuity of the probe is changed, when fracture is initiated at the contact, or both. A preliminary model for the physical origin of the extra crack-driving power of acute probes is presented in light of these, and complementary macro-scale *in-situ* indentation experiments. This work provides the basis for development of instrumented indentation-based nano-scale toughness measurement.

INTRODUCTION

Indentation by sharp diamond probes is a common procedure to investigate the elastic, plastic, viscous and fracture properties of materials in small volumes with little or no special sample preparation. Unfortunately, if fracture is to be studied, the frequently-used pyramidal Berkovich geometry rarely initiates fracture at loads common to nanoindentation, which are on the order of tens or hundreds of mN. The cube-corner indenter, a three-sided pyramid that is much more acute than the Berkovich, has been shown to be capable of reducing the load, and therefore the indentation length scale, needed to initiate radial fracture by several orders of magnitude [1, 2]. With knowledge of the elastic and plastic properties of the material, the crack lengths can be interpreted within the conventional model of indentation fracture to estimate the material toughness; for example, see the review by Cook and Pharr [3].

Essentially, the conventional model of indentation fracture asserts that constraint of the localized, irreversible plastic deformation zone causes an elastic reaction and accompanying tensile hoop stress at the material surface, which drives radial cracks. The reversible (not plastic) effect of indentation is to apply surface pressures; which, in classic indentation solutions such as those associated with Boussinesq, Love and Hertz, retard radial cracking with compressive surface hoop stresses outside the contacted zone. This elastic indentation + elastic-plastic model is sufficient to explain *in situ* transmission optical microscopy studies of crack growth during the indentation cycle for Vickers indentation, *e.g.* [4, 5].

Measurement of crack lengths at cube-corner indentations [2] *after* the indentation event seemed to confirm that the conventional elastic-plastic model of indentation fracture was obeyed when the indenter acuity was varied. That is, the crack lengths and the manner in which they scale with indentation load varied according to the scaling laws used in the conventional analysis. In fact, the cube-corner geometry appeared to be perform even better than Vickers or

Berkovich geometries when making toughness estimates; cube-corner indentation on the anomalous glasses fused silica and borosilicate glass obeyed the conventional indentation fracture model, whereas the fracture toughness of anomalous glasses is greatly overestimated from Vickers indentation [2]. The anomalous property of fused silica and borosilicate glass is their significant free volume, which allows them to compact underneath the pressures of indentation rather than plastically flow in a volume-conserving manner [6]. The competing deformation mechanism of compaction reduces the strength of the irreversible elastic-plastic field.

However, *in-situ* observation of radial crack growth during cube-corner indentation revealed that adherence to the conventional model was only coincidental [1]. The surprising effect of the reversible elastic field was to simultaneously drive and resist radial fracture, causing most crack growth to occur during loading with a weak contribution from the irreversible elastic-plastic field. This explains why it is possible to make good estimates of fracture toughness on the anomalous glasses with a cube-corner; plastic deformation is largely irrelevant to crack development. Later, a fracture-mechanics based indentation model described the reversible action of an acute indenter as a combination of a classic pressure-derived field and a crack-wedging field [7]. This model was successful at explaining much of the cube-corner indentation fracture phenomena [8], but the *ad hoc* mechanism of indentation wedging described there is unsatisfactory. In this work a potential explanation of indentation wedging phenomena is offered by appeal to experimental results from instrumented nanoindentation. The experiments are described below.

EXPERIMENTAL RESULTS

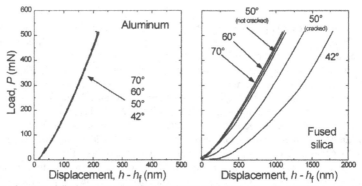

Figure 1. Load as a function of displacement minus final displacement (h-h_f) for four sharp probes on aluminum (left) and fused silica (right). Reproduced from Ref. [9].

The first set of experiments were reported in Morris *et al.* [9]. Four sharp diamond probe indenters were used; the Berkovich, the cube corner, and two 3-sided pyramids of intermediate acuity. The indenters are not referred to by their common names, but by the approximate half-included angle, ϕ, of the axisymmetric equivalent cone. Thus the Berkovich is 70°, the cube-corner 42°, etc. The materials indented were fused silica and soda-lime glasses, and single crystals of NaCl, BaF$_2$, Al, and Fe – 3% mass fraction Si. All crystals were $\langle 100 \rangle$-oriented. Indentations were all performed to peak loads of (50, 100, 200, 500 and 700) mN with an indentation strain rate of 0.050 s^{-1} on loading, a 30 s hold at peak load, then unloaded at a constant rate such that 90 % unload was completed in 45 s, and a thermal drift rate calculated before complete unloading. Scanning electron microscopy (SEM) was used to image the residual impressions after indentation.

The relevant results of [9] are quoted here. As expected, the more acute indenters produced more damage at fixed peak load, manifested simultaneously as irreversible work within the load-unload curve, increased pile-up, and fracture at the indentation site. More interestingly, for the three materials (Al, Fe and NaCl) that did not crack, the unloading traces are nearly perfectly superposable. This is shown in Figure 1 for aluminum. When there was radial fracture at the indentation site *and* the indenter was sufficiently acute (50° or 42°) then there was a distortion of the unloading curves away from the obtuse (70° or 60°) indenters. If there was a chance that a particular load-indenter-material combination would or would not fracture, then the unloading curves would either be distorted, or superpose upon Berkovich unloading, depending on whether or not fracture initiated during loading.

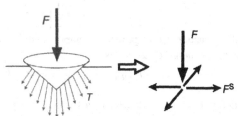

Figure 2. Resolved force F at a rigid conical punch in elastic-plastic indentation results in a traction distribution T that generally will have components normal and tangential to the undeformed surface. The tractions resolve to F and a pair of double forces F^S applied to the half space.

If the experimental results from the instrumented indentation studies are assumed to be widely applicable to homogeneous materials, then the following assertions are made. First, the elastic unloading traces from sharp indentation on uncracked materials are superposable, regardless of indenter acuity. From this, it is concluded that the elastic work done by the probe is (at least for the range of acuity in this experiment) acuity-invariant. Second, radial fracture at the indentation site is a necessary, but insufficient, condition for a compliant distortion of the unloading curve. This suggests that when the indenting probe is sufficiently acute, the effect of radial cracking is to lower the potential energy of the indenting probe; which, by reciprocity, indicates that the elastic stress field of the punch is driving the radial fracture. This is supported by *in-situ* experiments [1] that also show that there is a strong, reversible (and therefore not due to the reaction caused by the plastic deformation) stress field during cube-corner indentation that is weak for the Vickers indenter geometry.

DISCUSSION

Figure 2 is a schematic of elastic-plastic indentation with a sharp indenter. In general the tractions at the contact will not be normal to the undeformed surface. As shown on the right of Figure 2, the traction distributions will resolve in the far-field of the indentation as a point load F and a pair of surface shear double-forces of magnitude F^S. The full displacement and stress fields are known for these simple solutions and are listed in Ref [10]. In fact, the stress field of the double-forces of magnitude F^S is the surface-located blister field [10] and is obtainable by a simple differentiation of the Boussinesq point force solution [11].

A simple model is described below, with the intention of modeling the *far-field* stresses of the indentation as the acuity of the sharp probe increases. By this it is meant that the real boundary conditions at the contact (adhered, finite Coulomb friction or frictionless) are not meant to be captured. While the full stress and displacement fields are known for the point-force solutions shown in Figure 2, the energies are unbounded, and so cannot be used in the energy-balance scheme that follows.

Suppose that there are two individual elastic solutions that produce surface displacement fields u and u' by application, respectively, of surface tractions T_i and T_i' over a common part of the surface A. The work required to apply each displacement field individually is denoted by $U(u)$ or $U(u')$; for example,

$$U(u) = \frac{1}{2}\int_A T_i u_i dA. \tag{1}$$

Then, if the tractions of the primed and unprimed displacement fields act simultaneously on A, the total strain energy $U(u+u')$ is

$$U(u+u') = U(u) + U(u') + \int_A T_i' u_i dA. \tag{2}$$

A solution is desired such that $U(u+u')$ is unchanged from the reference state $U(u)$. For this to be true,

$$U(u') = \frac{1}{2}\int_A T_i' u_i' dA = -\int_A T_i' u_i dA = -\int_A T_i u_i' dA, \tag{3}$$

where the Betti reciprocal theorem has been invoked to form the last term on the right. If unprimed tractions are identified with those corresponding to a frictionless indentation solution, and primed tractions with frictionless axisymmetric outward shear, then Equation (3) forms a basis by which shear tractions may be coupled to normal pressure without changing the elastic work done on the body by the indenting punch. Frictionless pressure and shear solutions are not really necessary, but orthogonal T_i and T_i' simplify the analysis and retain the essential physics.

The reference indentation solution used is isotropic Hertzian contact. Cylindrical coordinates (z, r, θ) are used with the half-space occupying $z \geq 0$. Within the contacted circle of radius a, the Hertzian surface tractions T^H are a function of normalized radial distance $\rho = r/a$,

$$T^H = \frac{3F}{2\pi a^2}\left(1-\rho^2\right)^{1/2}, \tag{4}$$

where F is the total force acting on the punch. The T^H produce surface displacements w^H normal to the surface and u^H tangential to the surface. Now, an axisymmetric shear distribution T^S must be chosen to interact with the pressure distribution. Many solutions of this type are available, *e.g.* Ref. [11]; unfortunately, many of the published shear distribution solutions have unbounded energy (typically caused by finite tractions working through infinite displacements). In this work, a new surface shear distribution T^S is formed,

$$T^S \propto T^H \frac{1}{a}\frac{\partial}{\partial \rho}\left(w^H\right). \tag{5}$$

This form of T^S will fall smoothly to zero at the axis of indentation and at the edge of the contact area. When T^S is added to T^H in exactly the proportion of Equation (5), this will produce a

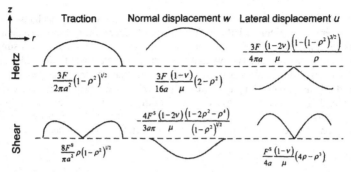

Figure 3. Tractions and displacements for $\rho \leq 1$ of the Hertzian solution and shear field of Equation (5). The dashed lines indicate zero.

distribution of surface traction $(T_i + T_i')$ that is normal everywhere to the normal Hertzian displacement field, w^H. From Equations (4) and (5), w and u for the shear tractions are found using the method of Noble and Spence [12]. These, along with the Hertzian tractions and displacements, are listed in Figure 3 as formulas and schematics for easy visualization. The shear force F^S is the vector sum of T^S over π radians, exactly as shown in Figure 2.

In general, addition of the shear tractions to the Hertzian tractions will change the strain energy of the body; but, because each field produces displacements in opposite directions (see Figure 3), the strain energy at fixed load can either be higher, lower or equal to that of the Hertzian field. Equation (3) is a formula that lets one find the magnitude of the shear field that will leave the strain energy unchanged. Substitution of the stresses and displacements in Figure 3 into Equation (3) with Hertzian indentation as the reference solution leads to

$$F^S = \frac{210}{256}\frac{(1-2v)}{(1-v)} F .$$ (6)

Therefore the shear tractions of Equation (5) may be added to the Hertzian field as defined by Equation (6) without changing the strain energy of the body or the force applied to the indenter. The shear force is proportional to the normal force, and is independent of the effective acuity of the indenter. For fixed traction distribution shapes, and fixed interaction surface area (both on the whole of A), this self-similar result is to be expected. Future work will attempt to model indenter acuity effects by breaking the similarity between the shear and pressure fields. There is no energetic difference between frictionless and perfectly adhered indentation for incompressible ($v = 1/2$) materials [13], and Equation (6) reflects this. Radial cracking then is analogous to the effect of slip at the punch face; crack formation relieves shear tractions and leads to punch settlement just as sudden deadhesion at a perfectly bonded punch face would.

CONCLUSIONS

Experiments of nanoindentation by four sharp diamond probes on a set of model materials, ranging in acuity from the Berkovich to the cube-corner, are reexamined. One remarkable result of that study is that elastic unloading load-displacement traces from a common peak load are superposable, except for when there is radial cracking at the indentation site *and*

115

the indenter is sufficiently acute. Experimental evidence for the wedging mechanism of radial crack development at cube-corner indentations, and not Vickers (or Berkovich) indentations, show that the reversible stress field of the indenting punch changes with acuity. A simple energetic analysis shows that acuity-invariant strain energy and the acuity-dependent wedging effect may be reconciled by adding outward axisymmetric shear to the pressures of indentation in certain proportions. Radial cracking at sufficiently acute indenters will release the forces of surface shear, and this is manifested as a distortion of the unloading curve away from that of an obtuse (*e.g.* Berkovich) indenter. Future work will focus on smoothly capturing the effect of indenter acuity on the stress fields during the entire indentation load-unload cycle.

REFERENCES

[1]   D. J. Morris and R. F. Cook, J. Am. Cer. Soc. 87 (2004) 1494.
[2]   G. M. Pharr, D. S. Harding and W. C. Oliver, in NATO ASI - Mechanical Properties and Deformation Behavior of Materials Having Ultra-Fine Microstructures, edited by M. Nastasi, D. M. Parkin and H. Gleiter (NATO ASI, 1993) p. 449.
[3]   R. F. Cook and G. M. Pharr, J. Am. Cer. Soc. 73 (1990) 787.
[4]   B. R. Lawn, A. G. Evans and D. B. Marshall, J. Am. Cer. Soc. 63 (1980) 574.
[5]   R. F. Cook and E. G. Liniger, J. Am. Cer. Soc. 76 (1993) 1096.
[6]   A. Arora, D. B. Marshall, B. R. Lawn and M. V. Swain, J. Non-Cryst. Solids 31 (1979) 415.
[7]   D. J. Morris and R. F. Cook, Int. J. Frac. 136 (2005) 237.
[8]   D. J. Morris, A. M. Vodnick and R. F. Cook, Int. J. Frac. 136 (2005) 265.
[9]   D. J. Morris, S. B. Myers and R. F. Cook, J. Mater. Res. 19 (2004) 165.
[10]   E. H. Yoffe, Phil. Mag. A 46 (1982) 617.
[11]   E. H. Yoffe, Phil. Mag. A 54 (1986) 115.
[12]   B. Noble and D. A. Spence, Formulation of two-dimensional and axisymmetric problems for an elastic half-space. University of Wisconsin Mathematics Research Center Report #1089, Madison, Wisconsin, (1971).
[13]   R. T. Shield and C. A. Anderson, Zeitschrift für Angewandte Mathematik und Physik 17 (1966) 663.

Mater. Res. Soc. Symp. Proc. Vol. 1049 © 2008 Materials Research Society   1049-AA05-03

An algorithm to determine the plastic properties of materials based on the loading data in single sharp indentation

Akio Yonezu, Hiroyuki Hirakata, and Kohji Minoshima
Mechanical Engineering, Osaka University, 2-1 Yamadaoka Suita, Osaka, 565-0871, Japan

ABSTRACT

We proposed an improved method to determine the plastic properties of a bulk material with only one sharp indenter. This method uses the data of loading curvature in an indentation curve to obtain the solution of the plastic properties. When the method yields two solutions, the indentation unloading data is employed to select the unique one. To verify the effectiveness of the proposed method, the plastic properties of four engineering materials were estimated on an experimental basis.

INTRODUCTION

Indentation is a useful technique to extract mechanical properties of material at micro and nano scale. Many researchers have explored the approach to determine the material constants of plastic constitutive equation, namely "plastic properties", using instrumented indentation. Cheng et al. [1,2] proposed the dimensional analysis, which connects the parameters of indentation curve to the plastic properties. Extensive numerical calculation for indentation has been carried out to establish the dimensionless function to estimate the plastic properties of various materials [3-5]. For this purpose, they generally use indentation loading data, such as loading curvature, which is sensitive to the plastic properties. Since, Dao et al. [3] made a significant contribution that employs the idea of representative stress - strain [6] for the dimensionless function, we can obtain one representative point in the plastic region of the stress - strain curve. However, the plastic properties involve two independent material constants, thus another function with complete independence is required. Therefore, another indentation curve with a different angle of the indenter tip is required to obtain another representative point. These two points yield the plastic properties, when the material is assumed to obey the work hardening rule. This methodology is called dual indentation [4, 5].

Although dual indentation is well established and is reported to allow determination of the plastic properties accurately [4, 5], it has some disadvantages such as the impression must be made twice on the same specimen and separated by a practical distance. This makes it impossible to obtain accurate information on the limited area of interest. Therefore, some researchers proposed the indentation method with just one sharp indenter (called single indentation), to determine the plastic properties [3,7]. They directly use the dimensionless function of unloading curve, which is independent of the loading curve. However, this method may be difficult to determine accurately, since the unloading data is not sensitive to the plastic properties. Thus, it is required that another function with a parameter of loading curve is employed to improve the single indentation.

In this study, we use different dimensionless functions from representative stress in order to improve the method of single sharp indentation. This method possesses a feature that the plastic properties are mainly estimated from the only loading curvature. We applied the proposed method to the four engineering materials on an experimental basis, to verify its effectiveness.

GENERAL METHOD

This study focuses on the material, which obeys the work hardening rule. The constitutive equation for the plastic region is expressed by yield stress σ_y, work-hardening exponent n and work-hardening strength R. When Young's modulus E is considered, R can be expressed by $\sigma_y(E/\sigma_y)^n$. Then, the plastic properties can be determined by σ_y and n, if E is known prior.

In an indentation curve of the relationship between indentation force F and penetration depth h, F can be expressed by Ch^2 (Kick's rule), when a sharp diamond indenter penetrates into homogenous solid. Then, C is independent of h, if the strain gradient effect is ignored. Based on dimensionless function [1,3], C is normalized by

$$\frac{C}{\sigma_r} = \Pi\left(\frac{E^*}{\sigma_r}, n\right) \cong \Pi\left(\frac{E^*}{\sigma_r}\right) \tag{1}$$

for a given indenter apex angle. Here, E^* represents the reduced elastic modulus. Dao et al. [3] proposed an interesting definition of representative stress σ_r and strain ε_r for the function of Eq.(1). This indicated that the function becomes independent of n, when the proper ε_r is selected.

In the unloading part, contact stiffness S and final penetration depth h_f are possible parameters for independent function of Eq.(1). It is usually difficult to measure h_f, since the value is affected by specimen roughness. Thus, S is used for the dimensional analysis [1,2]. Normalized S by maximum penetration depth h_{max} and E^* can be expressed by

$$\frac{S}{E^* h_{max}} = \Pi\left(\frac{\sigma_y}{E^*}, n\right) \tag{2}$$

In Eq.(2), previous research [3,7] used the representative stress σ_r instead of σ_y. Unique n can be identified, by substituting σ_r (from Eq.(1)) and experimental data (S, h_{max}) into Eq.(2). Then, we can basically determine the plastic properties from the values of n and σ_r. Here, the coefficients of polynominal Eqs.(1) and (2) are presented in Refs.[3,7].

PROPOSED METHOD

The relationship between the loading data C and the representative stress σ_r is already used for dimensionless function of Eq.(1). Therefore, an independent function of C - σ_r should be established to identify the plastic properties. Cheng et al. [1] proposed another dimensionless function, which is related to C. This uses the effective yield stress $(\sigma_y R)^{0.5}$ defined as σ_y^*. To establish this function, finite element method (FEM) calculations were carried out using the commercial code MARC (MSC software 2005r3, 62bit). A two dimensional axisymmetric model is created to simplify the calculation. The indenter is modeled as a rigid cone with a half apex angle of $70.3°$ to simulate Berkovich and Vickers indenter. The material properties were varied over a large range to cover essentially all engineering materials (E = 100 - 300 GPa, σ_y = 0.1 - 5 GPa, n = 0 - 0.5). Thus, 72 materials with different combinations of elasto-plastic properties were undertaken for the FEM calculation. The Poisson's ratio and friction coefficient are fixed at 0.3 and 0.15, which are minor factors for indentation [7].

Figure 1 shows the relationship between C/E^* and σ_y^*/E^*. All values are found to be independent of work hardening exponent n. Although the correlation factor $\mathcal{R}$ is relatively small ($\mathcal{R}=0.996$), the fitting function becomes

$$\frac{C}{E^*} = \Pi\left(\frac{\sigma_y^*}{E^*},n\right) \cong \Pi\left(\frac{\sigma_y^*}{E^*}\right) = -0.00342\left(\frac{\sigma_y^*}{E^*}\right)^4 - 0.07232\left(\frac{\sigma_y^*}{E^*}\right)^3 - 0.48713\left(\frac{\sigma_y^*}{E^*}\right)^2 - 0.79071\left(\frac{\sigma_y^*}{E^*}\right) + 1.844 \quad (3)$$

In order to evaluate the sensitivity of Eq.(3), reverse analysis was attempted. C values from FEM are substituted into Eq.(3) to estimate the effective yield stress σ_y^* for 72 materials. Figure 2 shows the error of σ_y^*, computed as $(\sigma_y^* \text{ rev.analysis} - \sigma_y^* \text{ input})/\sigma_y^* \text{ input}$, as a function of σ_y/E. The regions of $\sigma_y/E < 5 \times 10^{-4}$ and $\sigma_y/E > 1 \times 10^{-2}$ exhibited a large margin of error, up to 30 %. However, the middle region of $6 \times 10^{-4} < \sigma_y/E < 6 \times 10^{-3}$ showed a relatively small margin of error (mostly less than 10%). As shown by the two vertical lines in Fig.2, this range covers the properties of most engineering materials (see Ref. [8]). This suggested that Eq.(3) function can be used for engineering materials.

From Eqs.(1) and (3), C obtained from the loading curve is found to relate to two different plastic properties, σ_r and σ_y^*. With this idea, we propose a method to evaluate the plastic properties as shown in Fig.3. C from an indentation curve is substituted into two different functions. One is general function [3] of Eq.(1) to identify the representative stress σ_r. The other yields the effective yield stress σ_y^* based on Eq.(3). We then obtain two kinds of equations for work hardening exponent n as a function of σ_y. These were denoted as Equation A and B, suggesting that the unique n for σ_y can be obtained.

Reversed analysis was now carried out using numerical data of FEM calculation to verify the effectiveness of the proposed method. The proposed method was applied to two given materials with different plastic properties (E [GPa], σ_y [GPa], n) of (200, 0.1, 0.1) and (200, 1.0, 0.3). Here, Young's modulus E is known prior, since our purpose is to estimate the plastic properties. One indentation test was performed in the numerical experiment, then C was obtained from the indentation curve which solved the representative stress σ_r and the effective yield stress σ_y^*. As shown in Fig.3, these two approaches provide the two different functions of n for σ_y (Eqs. A and B). Figure 4 shows the change of n as a function of σ_y obtained from Eqs. A and B. In

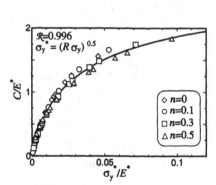

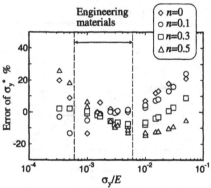

Figure 1 Dimensionless function of normalized C/E^* and σ_y^*/E^*.

Figure 2 Error of σ_y^* calculated by reversed analysis using dimensionless function of Fig.1.

Fig.4(a), one crossing point (0.1, 0.099) of two curves was found. This can be regarded to be the unique solution (σ_y, n). Contrary to this, two crossing points were found in Fig.4 (b), indicating that it is not always possible for the proposed method to yield a unique solution.

To solve this problem, we investigated the relationship between indentation curve and stress-strain curve obtained from the two solutions. Figure 5 shows the indentation curves (a) and stress-strain curves (b) calculated by the solutions, corresponding to the results in Fig.4 (b). Although the proposed method identified two different plastic properties, their indentation curves, especially loading curvature, were found to be the same. This indicated that the stress - strain curve that exhibits the same stress at the plastic strain of 3% (representative strain ε_r) gives the same loading curvature C (see Fig.5(a),(b)). In some materials, the proposed method is found to reduce the number of solutions to two, suggesting that the unique solution cannot be obtained from only one loading curvature C.

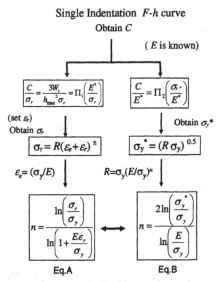

Figure 3 Proposed algorithm to determine plastic properties based on proposed single indentation.

We, then, focused on the dimensionless function of unloading data S [3].

For the result of Fig.5 (a), normalized S (= $S/E^*/h_{max}$) of 5.434 was obtained from the numerical experiment as indicated by the curve. We also calculated the normalized S from the two estimated solutions (denoted as solution 1 and 2 in Fig.4 (b)), based on Eq.(2)[3]. The experimental S (=5.434) is found to be closer to one (=5.528) of solution 2 than that (=4.967) of

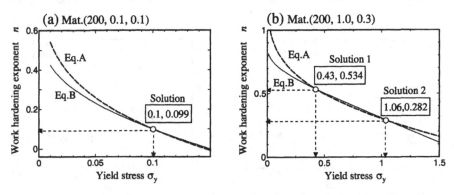

Figure 4 Changes of work hardening exponent n as a function of σ_y obtained from Equation A and B of Fig.3. Results of given two materials with (E, σ_y, n) were shown in (a),(b).

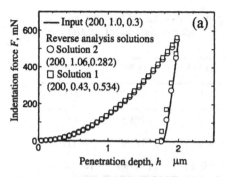 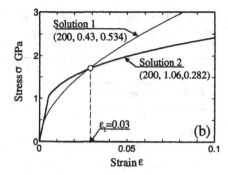

Figure 5 Indentation (a) and stress-strain curve (b) calculated by the two solutions obtained from Fig.4 (b). The elasto-plastic properties are shown as (E, σ_y, n).

solution 1. It is noted that the plastic properties (200, 1.06, 0.282) of solution 2 are close to the input properties (200, 1.0, 0.3). Therefore, the experimental S value is found to be closer to one calculated from the proper solution. This indicates that the usage of normalized S is useful to assess the unique solution, if the proposed method gives two solutions. It should be noted that this method uses only loading curvature to estimate the plastic properties. As well, this is a completely different method from the reported ones [3,7] where unloading data S is directly employed to determine the properties.

APPLICATION OF THE METHOD

The proposed single indentation was applied to determine the plastic properties of the four engineering metals, brass (C2801), alloy tool steel (SKD11), austenitic stainless steel (SUS304) and low carbon steel (SS400). Their elasto-plastic properties are shown in Table 1. Here, the material properties of SKD11 and SS400 are referred to Ref.[8] and the others were obtained from the uniaxial tensile tests.

Indentation tests with a Berkovich indenter were conducted using a commercial testing machine (Dynamic Ultra Micro Hardness Tester, DUH-201W: Shimadzu Co.). We tested more than 5 times, and these average data of loading curvature C and contact stiffness S and maximum penetration depth h_{max} were used for the proposed method. Here, we used Young's modulus E obtained from the tensile test, since the present purpose is to estimate the plastic properties. Table 1 shows the estimated plastic properties, indicating that the good agreements are found in the four materials. The maximum error of σ_y is 13 % in brass. The reported method [3] was also applied to the same materials to estimate the plastic properties as shown in Table 1. It was found that the results of brass and SS400 show good estimates, but those of SUS304 and SKD11 exhibit extremely large errors. This result is not clear, but is considered to be attributed in the error of the fitting functions for unloading data [3]. Among the four engineering materials in Table 1, it was shown that their plastic properties could be evaluated with the proposed method using two dimensionless functions of loading curvature, which was sensitive to the plastic properties.

Table 1 The plastic properties of four engineering materials obtained from the uniaxial tensile test. The predicted properties by proposed and reported method [3] are also shown. The error values are given by (%).

		Brass	SKD 11[8]	SUS 304	SS 400[8]
E GPa	Tensile test	103	216	196	211
σ_y MPa	Tensile test	120	243	203	272
	Proposed method	104 (13%)	220 (9%)	209 (3%)	241 (11%)
	Reported method [3]	91 (24%)	41 (83%)	21 (89%)	161 (40%)
n	Tensile test	0.27	0.276	0.3	0.235
	Proposed method	0.307 (13%)	0.307 (11%)	0.27 (9%)	0.244 (4%)
	Reported method [3]	0.34 (24%)	0.545 (98%)	0.61 (103%)	0.33 (40%)

CONCLUSIONS

We proposed an improved method to determine the plastic properties of a bulk material, based on the data of loading curvature in an indentation curve with only one sharp indenter. Two kinds of dimensional functions for loading curvature were employed to identify the representative stress and the effective yield stress. If these functions yield two solutions, an indentation unloading data (contact stiffness, S) is employed to select the unique one among the two. This method has the potential to effectively evaluate the plastic properties, since the properties of the four engineering materials could be determined successfully

ACKNOWLEDGMENTS

The work of A.Y. and H.H. is supported in part by Grant-in-Aid for Young Scientist of (B) (No.19760075 and 187600830) of the Ministry of Education, Culture, Sports, Science and Technology, Japan.

REFERENCES

1. Y-T Cheng and C-M Cheng, Journal of Applied Physics, Vol.84, pp.1284–1291, (1998).
2. Y-T Cheng and C-M Cheng, Materials Science and Engineering, Vol.44, pp.91–149, (2004).
3. M. Dao, N. Chollacoop, K. J. Van Vliet, T. A. Venkatesh and S. Suresh, Acta Materialia, 49, pp.3899-3918 (2001).
4. J. L. Bucaille, S. Stauss, E. Felder and J. Michler, Acta Materialia, Vol.51, pp.1663-1678 (2003).
5. X. Chen, N. Ogasawara, M. Zhao and N. Chiba, Journal of Mechanical Physics of Solids, Vol.55-8, pp. 1618-1660 (2007)
6. D. Tabor, "The Hardness of Metals", Oxford Classic Texts, (1950)
7. N. Ogasawara, N. Chiba, Xi Chen, Scripta Materialia, Vol.54, pp.65-70 (2006).
8. S.H. Kin, B.W. Lee, Y. Choi and D. Kwon, Materials Science & Engineering A, Vol.415, pp.59-65 (2006).

Mater. Res. Soc. Symp. Proc. Vol. 1049 © 2008 Materials Research Society

PLASTICITY CHARACTERISTICS OBTAINED THROUGH INSTRUMENTAL INDENTATION

Yuliy Milman[1], Sergey Dub[2], and Alex Golubenko[1]

[1]Physics of High-Strength and Metastable Alloys, Institute for Problems of Material Science, 3, Krzhizhanovski Str., Kiev, 03142, Ukraine

[2]Mechanical Properties of Superhard Materials, Institute for Superhard Materials, 2, Avtozavodskaya Str., Kiev, 04074, Ukraine

ABSTRACT

The nanohardness and microhardness testing of crystalline materials with different types of interatomic bonds and different crystal structures was performed with a Berkovich indenter.

The plasticity characteristics was determined from the ratio of area under the unloading curve to the area under the loading curve (δ_A) in the process of instrumented indentation. It is shown that δ_A coincides with the plasticity characteristic δ_H calculated from Young's modulus E, microhardness HM, and Poisson's ratio ν at $\delta_H \geq 0.5$ according to a theory developed earlier, i.e., for all metals and most covalent crystals and ceramics.

A relation describing the dependence of the plasticity characteristic on the load applied to an indenter, provided that the Meyer relation is satisfied, was obtained.

It is experimentally shown that at $\delta_A > 0.5$, the plasticity characteristic δ_A is determined by the expression $\delta_A = 1 - 10.2 \cdot HM(1 - \nu - 2\nu^2)/E$, which was obtained theoretically for δ_H early.

Plasticity characteristics δ_A and δ_H may be used for characterization of mechanical behaviour of materials which are brittle at standard mechanical tests and for coatings.

INTRODUCTION

The majority of modern materials (excluding metals and metallic alloys) are low-plasticity or even brittle materials in standard mechanical tests.

However, the plasticity of these materials can be characterized by the indentation method. Since plasticity is usually characterized as a capability of a material to retain a part of strain, caused by loading, after unloading, it was proposed [1] to characterize plasticity with a dimensionless parameter equal to the fraction of plastic strain in the total elastoplastic strain

$$\delta = \varepsilon_p / \varepsilon_t = \varepsilon_p / (\varepsilon_p + \varepsilon_e) = 1 - \varepsilon_e / \varepsilon_t, \tag{1}$$

where ε_p, ε_e and ε_t are, respectively, the plastic strain, elastic strain, and total strain, at that, $\varepsilon_t = \varepsilon_p + \varepsilon_e$.

Plasticity characteristic obtained by indentation may be used for characterization of mechanical behaviour of materials which are brittle at standard mechanical tests, for coatings and thin layers.

The average elastic strain on an indenter – specimen contact area in the direction of the load applied to the indenter was obtained in the form [1]

$$\varepsilon_e = -HM(1-\nu-2\nu^2)/E, \qquad (2)$$

Here HM is the Meyer hardness, which is considered as an average contact pressure, E is Young's modulus, and ν is Poisson's ratio of the investigated material.

The total strain ε_t was determined for pyramidal indenters in the form: $\varepsilon_t = -ln\sin\gamma$, where γ is the angle between a face and the axis of the pyramid.

Then, according to (1), the plasticity characteristic determined in indentation for Berkovich indenter ($\gamma = 65°$) is as follows:

$$\delta_H = 1 + HM(1-\nu-2\nu^2)/E \cdot ln\sin\gamma = 1 - 10.2 \cdot HM(1-\nu-2\nu^2)/E. \qquad (3)$$

As can be seen from (1), the relation $0 < \delta_H < 1$ must be satisfied. It was shown that, for materials plastic in tension, $\delta_H \geq 0.9$. It was also shown that there exists a correlation between the value δ_H and the elongation to fracture in tensile tests [1].

The theory of determination of the plasticity characteristic δ_H was further developed in [2]. In this work, the condition of incompressibility of a material under an indenter was used to calculate only the plastic component of strain ε_p but not the total strain, as it was done in [1]. That is why results obtained in [2] can be used in the computation of strains and the plasticity characteristic δ_H for hard and superhard materials with a large fraction of the elastic strain in indentation and in the indentation with a blunter indenters with a larger value of the angle γ.

For the plastic strain, the following relation was obtained for pyramidal indenters:

$$\varepsilon_p = -ln\left[1 + \left(ctg\gamma - HM/0.565 \cdot E_{ef}\right)^2\right]^{1/2}. \qquad (4)$$

Here E_{ef} is effective Young's modulus of the contact pair indenter – specimen [2].

Using relations (4), (2), and (1), the plasticity characteristic δ_H is calculated.

In [2, 3], it was shown that, for metals, such a calculation gives values of δ_H coinciding with results obtained with (3). Differences are observed only for hard and superhard materials at $\delta_H < 0.4$, and for these materials calculations should be performed using relations (4), (2), and (1).

In the present work, for a number of materials, the calculation of δ_H was performed using (4), (2), and (1), but in the discussion of the dependence of δ_H on the ratio HM/E we will use Eq. (3), which is more convenient for this purpose.

In instrumented indentation the plasticity characteristic can be determined as the ratio of the work of plastic deformation to the total work of deformation from the diagram presented in Fig. 1 in conformity with ISO 14577-1:2002(E)

$$\delta_A = A_p/A_t = A_p/(A_p + A_e) = 1 - A_e/A_t. \qquad (5)$$

In the present work, the plasticity characteristics δ_H and δ_A have been determined for a large number of crystals with different character of interatomic bonds and different crystal structure. The correlation of these characteristics is discussed.

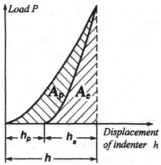

Figure 1. Diagram of penetration of a pyramidal indenter in the coordinates load P– displacement of the indenter h; h_e and h_p are the elastic and plastic approachment of the indenter and specimen to one another; A_e and A_p are the elastic and plastic component of the work of deformation in instrumented indentation.

EXPERIMENTAL DETAILS

Materials tested by indentation are presented in Table I. Mainly single crystals or polycrystalline materials of high purity were investigated. The instrumented indentation was carried out using a Nano Indenter II unit (MTS Systems, USA) with a diamond indenter in the form of a Berkovich trihedral pyramid. The nanohardness and Young's modulus were calculated by the Oliver–Pharr technique [4].

Table I. Microhardness, load on indenter, Young's modulus, and values of the plasticity parameters calculated by the nanoindentation and microindentation method and elongation to fracture in tension test (δ, %).

№	Materials	P, mN	E, GPa	ν	HM$_{(MICRO)}$, (GPa) P=200g	HM$_{(NANO)}$, (GPa) instrumented hardness	h, nm	h$_e$, nm	A= A$_e$+A$_p$	A$_e$	δ_A	$\delta_{H(NANO)}$	$\delta_{H(MICRO)}$ P=200g	δ, %
1	BeO*	10	400	0.23	10.85	12.8	181.5	43.3	739.2	179	0.76	0.77	0.81	0
2	Si₃N₄**	50	324	0.30	22.29	24.3	415.3	199.1	7826	3769	0.52	0.54	0.58	0
3	NbC*	50	550	0.21	29.45	31.3	359.3	158.5	6831	3105	0.54	0.55	0.58	0
4	ZrN*	50	400	0.25	20.74	24.3	400.7	161.6	7799	3258	0.58	0.57	0.64	0
5	WC*	50	700	0.31	25.72	39.8	310.6	110.9	6144	2303	0.62	0.62	0.78	0
6	β-SiC*	50	460	0.22	36.84	44.3	323.2	169.6	6055	3511	0.42	0.29	0.40	0
7	ZrC**	50	480	0.20	25.15	26.4	386.0	158.3	7319	3130	0.57	0.57	0.59	0
8	B₄C°	10	500	0.15	40.61	48.9	123.3	78.3	477	318	0.33	0.23	0.34	0
9	Si*	50	169	0.22	11.45	11.9	495.7	223.9	8958	4519	0.50	0.52	0.53	0
10	Ge*	30	130	0.21	9.43	10.7	473.2	216.6	5040	2474	0.51	0.44	0.50	0
11	W*	10	420	0.28	3.72	6.10	301.3	33.9	1258	125	0.90	0.91	0.95	5
12	Mo*	50	324	0.29	2.23	3.21	931.2	55.2	18664	1244	0.93	0.94	0.96	10
13	Cr*	50	279	0.31	1.70	2.63	1025.3	55.2	20495	1256	0.94	0.95	0.97	12
14	Nb**	50	104	0.40	1.10	1.26	1460.2	87.1	27144	1973	0.93	0.96	0.97	20
15	Ta**	50	185	0.34	1.35	1.74	1259.2	60.9	24340	1349	0.94	0.96	0.97	22
16	Cu*	50	130	0.34	0.46	0.66	2223.1	52.3	41462	1417	0.97	0.98	0.98	40
17	Al**	120	70	0.35	0.36	0.66	3148.0	109.4	144435	5524	0.96	0.96	0.98	45
18	Ti**	50	120	0.36	2.10	2.85	1053.9	123.3	1712	187	0.89	0.90	0.93	6

* – single crystal,   ** – polycrystal,   ° – individual grain

The microhardness was tested using a PMT-3 unit and a Berkovich indenter. The load applied to the indenter was 2 N. This made it possible to study the scale dependences of the hardness and plasticity characteristic δ_H using the same type of indenters. It is seen from Table I that plasticity characteristics δ_H and δ_A can be determined for ductile metals and for brittle at tension tests materials as well.

DISCUSSION

Scale Dependence of the Plasticity Characteristic δ_H.

In Fig. 2, the values of the plasticity characteristic δ_H calculated from the results of instrumented indentation ($P = 10 - 120$ mN, hardness was calculated according to [4]) and from the results of microhardness tests ($P = 2$ N) are compared. In both cases, the calculations were performed using Eqs. (1), (2), and (4). From Table I it is seen that the load in microindentation was 20 or even 200 times larger than that in nanoindentation.

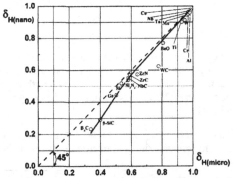

Figure 2. Dependence of the plasticity characteristic obtained in nanoindentation ($\delta_{H(nano)}$, $P = 10 - 120$ mN, calculation of hardness and Young's modulus by [4]) on the value of the plasticity characteristic obtained in microindentation ($\delta_{H(micro)}$, $P = 2$ N) for different materials.

The size dependence of hardness was discussed in [5]. Dislocation model for investigation of the size effect of hardness was proposed in [6] and was developed in [7] and [8]. The homogeneous deformation during indentation is considered in these works with use of concepts of geometrically necessary and stored dislocations.

In the present work we have obtained some results for size dependence of plasticity characteristic δ_H. From Fig. 2 it follows that $\delta_{H(micro)}$ is always slightly higher than $\delta_{H(nano)}$. The smaller values of these characteristics, the larger the difference between these. If $\delta_H > 0.5$, i.e., for metals and most ceramic materials, the difference between these values is insignificant.

Using results published in [5] for size dependence of hardness we get the relation

$$\left(1 - \delta_{H_1}\right) / \left(1 - \delta_{H_2}\right) = \left(P_1 / P_2\right)^{1-2/n}, \qquad (6)$$

where δ_{H1} and δ_{H2} are the plasticity characteristics under loads P_1 and P_2, n is the exponent in the Meyer relation $P = \text{const } h^n$ (the value of n see in [5]). Obtained results (see Table I) show that relation (6) is satisfied satisfactorily for metals, but, it is practically not satisfied for hard ceramic materials.

At the same time, Fig. 2 can be used as a calibration curve for the recalculation of $\delta_{H(nano)}$ to $\delta_{H(micro)}$, and vice versa.

Relation between the Plasticity Characteristics δ_H and δ_A and the Value of $HM(1-v-2v^2)/E$.

Expression (3) predicts a linear dependence of δ_H on $HM(1-v-2v^2)/E$. From Fig. 3 it follows that relation (3) is satisfied well for δ_H and δ_A at values of these characteristics above 0.5. In this case, the slope of the straight line α in Fig. 3 coincides well with the theoretical value of $tg\alpha = 1/\ln\sin\gamma \approx 10.2$, which follows from (3). Thus, the experimental confirmation of expression (3), obtained in [1], and the deviation from it at small values of δ_H, predicted in [2], were shown.

From Fig. 3 it is seen also a rather satisfactory coincidence of the values δ_H calculated by formulas (1), (2), (4) and the values of δ_A determined from the ratio of the area under the unloading curve to the area under the loading curve when both values were determined at the same load.

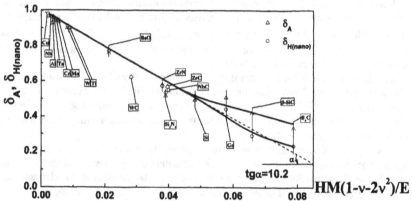

Figure 3. Dependences of the plasticity characteristic δ_A (calculated from the ratio of the area under the unloading curve to the area under the loading curve) and the plasticity characteristic $\delta_{H(nano)}$ (calculated from the value of the nanohardness by formulas (1), (2), and (4)) on the ratio $HM(1-v-2v^2)/E$ for different materials.

It should be recognized that both methods of determination of the plasticity characteristic can be considered to be equally applicable at $\delta_H > 0.5$, i.e., for all metals and most refractory compounds.

Let us represent the total work on penetration of the indenter in the form $A_t = V\int_0^{\varepsilon_t} \sigma d\varepsilon$.

Here V is the volume of the indentation print, σ is the average acting stress on the contact area, and the integral corresponds to the work made by the indenter for the formation of unit volume of the indentation print. We will assume that $\sigma \cong HM = const$. Then $A_t = V \cdot HM \cdot \varepsilon_t$, and the work of elastic deformation is $A_e = V \cdot HM \cdot \varepsilon_{ek}$. Here ε_{ek} is the elastic strain of the indenter – specimen contact pair.

Using (5) we obtain

$$\delta_A = \varepsilon_p / (\varepsilon_p + \varepsilon_{ek}), \tag{7}$$

which is very close to the expression for the plasticity characteristic δ_H determined according to (1). The difference between expressions (7) and (1) is that, in contrast to ε_e in (1), the term ε_{ek} in (7) takes into account both the elastic strain of the indentation print (ε_e) and the elastic strain of the diamond indenter ($\varepsilon_{e\,ind}$), so that $\varepsilon_{ek} = \varepsilon_e + \varepsilon_{e\,ind}$.

Taking into consideration the results presented above ($\delta_A \approx \delta_H$), it should be recognized that, for materials with $\delta_H > 0.5$, the contribution of the work for the elastic deformation of the diamond indenter to the total work of deformation is insignificant due to the high Young's modulus of diamond. For very brittle materials only ($\delta_H < 0.4$), the fact that $\delta_A > \delta_{H(nano)}$ that may be the consequence of fracture in the indentation print and decreasing the ε_e in (7).

CONCLUSIONS

It is shown that the plasticity characteristic δ_A, obtained in instrumented indentation practically coincides with the plasticity characteristic δ_H, which is calculated for Berkovich indenter according to [1, 2] from the relation $\delta_H = 1 - 10.2 \cdot HM(1 - v - 2v^2)/E$. It is shown that a decrease in the load P leads to not only an increase in the hardness, but also a reduction in the plasticity if the Meyer relation is satisfied.

The obtained results indicate that it is expedient to carry out the determination of hardness and Young's modulus in instrumented indentation with the determination of the plasticity characteristic δ_A. This characteristic determines to a substantial degree the mechanical behavior of materials, which was shown early for the analogous characteristic δ_H, determined in measurements of the microhardness.

It is important that the determination of the plasticity characteristic δ_A in instrumented indentation does not require knowledge of other parameters of the material (hardness, Young's modulus, Poisson's ratio). For this reason the accuracy of determination of this characteristic in comparison with the accuracy of determination of the plasticity characteristic δ_H in microindentation.

REFERENCES

1. Yu.V.Milman, B.A.Galanov, S.I.Chugunova, *Acta Metall. Mater.* 41(9), 2523-2532 (1993).
2. B.A.Galanov, Yu.V.Milman, S.I.Chugunova, I.V.Goncharova, *Sverkhtverdye Materialy.* 3, 25-38 (1999).
3. A.V.Byakova, Yu.V.Milman, A.A.Vlasov, *J. Science of Sintering.* 36, 27-103 (2004).
4. W.C. Oliver and G.M. Pharr, *J. Mater. Res.*, 7, 1564-1583 (1992).
5. Yu.V.Milman, *Problemy Prochnosti*, 6, 52-56 (1990).
6. N.Stelmashenko, M.Walls, L.Brown and Yu.Milman, *Acta Metall. Mater.* 41(10), 2855-2865 (1993).
7. W.Nix, H.Gao, *J. Mechanics and Physics of Solids*, 46, 411-425 (1998).
8. F.R.N.Nabarro, S.Shrivastava, S.B.Luyckx, *Phil. Magazine*, 86(25-26), 4173-4180 (2006).

Mater. Res. Soc. Symp. Proc. Vol. 1049 © 2008 Materials Research Society   1049-AA05-01

Nanoindentation Analysis of Plasticity Evolution during Spherical Microindentation of Bulk Metallic Glass

Byung-Gil Yoo, and Jae-il Jang

Division of Materials Science and Engineering, Hanyang University, Seoul, Korea, Republic of

ABSTRACT

Unlike most of crystalline metals, metallic glasses are known to exhibit a fully-plastic behavior or work softening during mechanical deformation. To analyze the characteristics of the deformed region, here a series of instrumented micro- and nano-indentation experiments were performed on a Zr-based bulk metallic glass (BMG) with geometrically self-similar sharp indenter as well as spherical indenters. First, we performed instrumented micro-indentation tests with a spherical indenter on the bonded interfaces of the BMGs. Although adhesive (used for bonding the interfaces) might significantly affect the deformation mode by reducing the constraint, the evolution of subsurface plasticity during spherical indentation was clearly observed. Subsequently, the subsurface plasticity underneath the hardness impressions was systematically examined through nanoindentation. The results are discussed in terms of major change in mechanical responses of BMGs before and after indentation-induced deformation.

INTRODUCTION

It is well known that bulk metallic glasses (BMGs) show unique plastic deformation at low temperature and high stress, i.e., once plasticity is initiated, it is highly localized into narrow shear bands [1-2]. Recently, instrumented indentation technique [3-4] has been widely used for analyzing this interesting phenomenon of BMGs. In particular, a new focus of many indentation works on BMGs is their inhomogeneous plastic flow during indentation experiments [5]. A number of studies have reported reproducible 'serrations' in nanoindentation P-h curve and now it is well accepted that the serrations are associated with shear bands nucleation and/or propagation [6-10]. Nevertheless, there have been difficulties in assessing practical shear banding phenomena during indentation, simply because the shear bands are often captured beneath the indenter and cannot expand to a free surface [11]. To overcome the difficulties, 'interface bonding technique' has been extensively applied to explore the subsurface deformation of BMGs underneath the indentation impression [12-17]. However, previous studies have been mostly limited to the analysis of the subsurface deformation induced by Vickers indentation at different loads [12-14,16-17] or by spherical indentation at a specific given load [15]. Unlike indentation with a geometrically self-similar sharp indenter, indentation with a spherical indenter produces increase in representative stress and strain as the applied load is increased. Therefore, by analyzing the subsurface morphology of spherical indentation made at various loads, one might gain somewhat new insights on the inhomogeneous plastic deformation of BMGs. With this in mind, here we examined the evolution of plastic deformation in a Zr-based BMG during spherical indentation by performing both macroscopic instrumented indentation (with a spherical indenter) and nanoindentation (with a three-sided pyramidal Berkovich indenter).

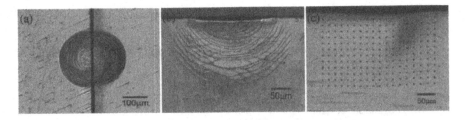

Figure 1. Optical micrographs demonstrating the testing procedure applied in this work: (a) macroscopic indentation on the bonded interface with a spherical indenter; (b) observation of subsurface deformation morphology; (c) nanoindentation after gentle polishing of the deformed region.

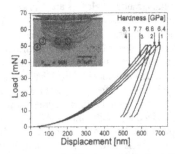

Figure 2. Representative P-h curve recorded during nanoindentation on different regions underneath the macroscopic indentation impression.

EXPERIMENTAL DETAILS

$Zr_{52.5}Cu_{17.9}Ni_{14.6}Al_{10}Ti_5$ BMGs for the interface-bonding technique were prepared by cutting the rod first into two halves and then polishing them to a mirror finish prior to bonding them using a high-strength adhesive (Loctite, Henkel Ireland Ltd., Ireland). As the adhesive's mechanical properties were not provided by its manufacturer, we have not quantitatively considered the effect of the adhesive. The top surface of the bonded specimen was polished so that it was flat like a mirror. On the bonded interface, macro-scale indentations were carried out using an instrumented indentation equipment, AIS-2100 (Frontics Inc., Seoul, Korea), with a WC spherical indenter having a radius of 500 μm [see figure 1(a)]. The maximum indentation loads were varied in the range from 19.6 to 196 N, and loading rate was fixed as 5 μm/s. After spherical indentation, the bonded interface was opened by dissolving the adhesive in acetone, and then the subsurface deformation morphology was observed through optical microscope [figure 1(b)]. Subsequently, the subsurface deformation zone was polished again into a flat surface using alumina particle of 0.3 μm. In order to evaluate the hardness distribution within the subsurface deformation region, a series of nanoindentation experiments were performed on the gently polished surface using a Nanoindenter-XP (MTS Corp., Oak Ridge, TN) [figure 1(c)]. The maximum indentation load and the loading rate were 50 mN and 0.5 s^{-1} respectively. After nanoindentation testing, the profiles of the indented surfaces were examined by atomic force microscopy (AFM) XE-100 (Park Systems, Suwon, Korea).

RESULTS AND DISCUSSION

Figure 2 shows a typical example of P-h curves recorded during indentations on different subsurface deformation regions. There is clear difference in the curve and thus hardness value; hardness values of the extensively deformed regions (region '1' and '2' in the figure) are significantly lower than those of the region near shear bands (region '3') and un-deformed region (region '4'). Note that the nanoindentation hardness values were calculated according to Oliver-Pharr method so that they are overestimated than the real value, as that cannot take into consideration the pile-up typically observed around the hardness impression of BMGs [18].

Nanoindentation hardness distribution for the region underneath indentation impression is mapped in figure 3 with a background of optical microscopy image for the subsurface deformation morphology. It was found that the shear bands zones were indeed softened although there was some fluctuation in the hardness value. As indentation load is increasing, the tendency for the change in the size of the softened zone (i.e., black triangle points in the figure, whose hardness is lower than 7 GPa) is in a good agreement with that for the variation in the shear bands zone size. The nanoindentation hardness value for the softened zone is plotted in figure 4 as a function of indentation load. High load indentations show slightly lower hardness values than low load indentation, which implies that the extent of softening keeps increasing as spherical indentation load is increasing. Very recently Bei et al. [19] reported in their paper on the same BMG as used here that the hardness of the compressed sample was continuously

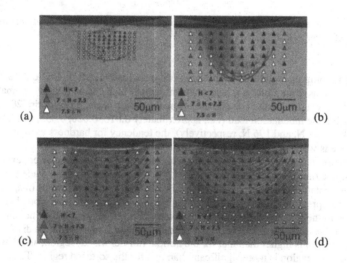

(a) (b) (c) (d)

Figure 3. Hardness distribution underneath the indentation made at various peak load: P_{max} = (a) 19.6 N, (b) 49 N, (c) 98 N, and (d) 196 N.

131

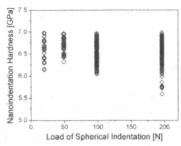

Figure 4. Variation of nanoindentation hardness with increase in the load of spherical indentation.

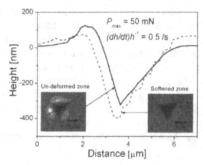

Figure 5. Typical example of AFM analysis of the hardness impressions made in softened region and in un-deformed region.

decreasing as compressive plastic strain increases in a wide range from 0% to 80%. Since representative plastic strain for a spherical indentation (which can be roughly estimated as 0.2(a/R) where a and R are the contact radius and the radius of the indenter tip [20]) is increasing with increasing the indentation load (i.e., approximately 0.016, 0.026, 0.035, and 0.049 for P_{max} = 19.6 N, 49 N, 98 N, and 196 N, respectively), the tendency for hardness vs. indentation load observed here is well matched with Bei et al.'s results [19].

Deformation behaviors of softened and normal (un-deformed) regions were examined more in detail using atomic force microscopy (AFM). Four measurements were made for each region and the representative example of the AFM work is shown in figure 5. AFM image taken from the hardness impression of the un-deformed region shows higher contrast around the indentation, which is due to the material pile-up, than that of the softened region. In the line scan profile across the hardness impression, it is clearly seen that (1) the final indentation displacement for the softened region is higher than that for the normal (un-deformed) region, and that (2) the pile-up for the normal region is more significant than that for the softened region. The same tendency was observed in all AFM measurements though there were some fluctuations in the pile-up amount. Somewhat interestingly, this observation is opposite to the results in previous study [15] where the softened region shows bigger pile-up than the un-deformed region. Severe material pile-up around indentation is a nature of a material without work (strain) hardening behavior; i.e., due to the incompressibility of the material, the material removed from the indented volume can pile up around the indentation. According to free volume model by Spaepen [21], considerable amount of excess free volume can be produced in the plastic zone under the indenter during the macroscopic spherical indentation. During subsequent nanoindentation, the produced excess free volume can enhance the ability to accommodate plastic deformation induced by nanoindentation. This might result in smaller pile-up in the deformed region than in un-deformed region.

CONCLUSION

In this paper, we reported our recent observation on the evolution of subsurface plastic deformation in a Zr-based BMG during spherical indentation. It was revealed that the subsurface region under the indenter was indeed softened and had quite different deformation characteristics from that of un-deformed region.

ACKNOWLEDGMENTS

This research was supported by the Korea Research Foundation Grant funded by the Korean Government, MOEHRD (Grant No. KRF-2006-331-D00273). The authors would like to thank Dr. H. Bei (at Oak Ridge National Laboratory) for providing the valuable sample.

REFERENCES

1. W. H. Wang, C. Dong, and C. H. Shek, *Mater. Sci. Eng. R* **44**, 45 (2004).
2. C. A. Schuh, T. C. Hufnagel, and U. Ramamurty, *Acta Mater.* **55**, 4067 (2007).
3. W. C. Oliver, and G. M. Pharr, *J. Mater. Res.* **7**, 1564 (1992).
4. W. C. Oliver, and G. M. Pharr, *J. Mater. Res.* **19**, 3 (2004).
5. C. A. Schuh, and T. G. Nieh, *J. Mater. Res.* **19**, 46 (2004).
6. C.A. Schuh, and T. G. Nieh, *Acta Mater.* **51**, 87 (2003).
7. C. A. Schuh, A. C. Lund, and T. G. Nieh, *Acta Mater.* **52**, 5879 (2004).
8. A. L. Greer, A. Castellero, S. V. Madge, I. T. Walker, and J. R. Wilde, *Mater. Sci. Eng. A* **375-377**, 1182 (2004).
9. B. C. Wei, L. C. Zhang, T. H. Zhang, D. M. Xing, J. Das and J. J. Eckert, *J. Mater. Res.* **22**, 258 (2007).
10. J.-I. Jang, B.-G. Yoo, and J.-Y. Kim, *Appl. Phys. Lett.* **90**, 211906 (2007).
11. B.-G. Yoo, J.-Y. Kim, and J.-I. Jang, *Mater. Trans.* **48**, 1765 (2007).
12. S. Jana, U. Ramamurty, K. Chattopadhyay, and Y. Kawamura, *Mater. Sci. Eng. A* **375-377**, 1191 (2004).
13. S. Jana, R. Bhowmick, Y. Kawamura, K. Chattopadhyay, and U. Ramamurty, *Intermetallics* **12**, 1097 (2004).
14. U. Ramamurty, S. Jana, Y. Kawamura, and K. Chattopadhyay, *Acta Mater.* **53**, 705 (2005)
15. R. Bhowmick, R. Raghavan, K. Chattopadhyay, U. Ramamurty, *Acta Mater.* **54**, 4221 (2006)

16. H. Zhang, X. Jing, G. Subhash, L. J. Kecskes, and R. J. Dowding, *Acta Mater.* **53**, 3849 (2005).
17. W. H. Li, T. H. Zhang, D. M. Xing, B. C. Wei, Y. R. Wang, and Y. D. Dong, *J. Mater. Res.* **21**, 75 (2006).
18. W. J. Wright, R. Saha, and W. D. Nix, *Mater. Trans., JIM* **42**, 642 (2001).
19. H. Bei, S. Xie and E. P. George, *Phys. Rev. Lett.* **96**, 105503 (2006).
20. D. Tabor, *The Hardness of Metals* (Oxford University Press, London, 1951).
21. F. Spaepen, *Acta Metall.* **25**, 407 (1977).

Mater. Res. Soc. Symp. Proc. Vol. 1049 © 2008 Materials Research Society   1049-AA05-08

Reverse Plasticity in Nanoindentation

Yongjiang Huang[1,2], Nursiani Indah Tjahyono[2], Jun Shen[1], and Yu Lung Chiu[2]

[1]School of Materials Science and Engineering, Harbin Institute of Technology, Harbin, 150001, China, People's Republic of

[2]Chemical and Materials Engineering, University of Auckland, 20 Symonds Street, Auckland, 1142, New Zealand

ABSTRACT

This paper summarises our recent cyclic nanoindentation experiment studies on a range of materials including single crystal and nanocrystalline copper, single crystal aluminium and bulk metallic glasses with different glass transition temperatures. The unloading and reloading processes of the nanoindentation curves have been analysed. The reverse plasticity will be discussed in the context of plastic deformation mechanisms involved. The effect of loading rates on the mechanical properties of materials upon cyclic loading will also be discussed.

INTRODUCTION

Nanoindentation technique has been widely used to study the mechanical properties of various materials such as modulus and hardness from the load-displacement curves for its rapidity, simpleness and precision over the last two decades [1].

Traditionally, the elastic modulus of a testing material can be derived from the unloading curve assuming that the unloading is an elastic recovery. However, it has been noted that the measurement on metallic samples may result in elastic moduli higher than those measured by tensile tests. This has been ascribed to the reverse plasticity during unloading where the internal stress developed during loading becomes unstable [2].

In addition to the singular loading-unloading indentation test, cyclic indentation can be carried out at either constant loading rate or constant displacement rate [3]. The cyclic indentation can be used to study quantitatively the hardening or softening behaviour of materials under cyclic loading conditions [4] as well as the fatigue performance of materials [5].

The present work is an attempt to study systematically the reverse plasticity in the unloading/reloading process where a large range of materials were subjected to cyclic indentation.

EXPERIMENT

The testing materials used for the present study include single crystal aluminium, single crystal copper, nanocrystalline copper, and titanium-based and iron-based bulk metallic glasses (BMG). The single crystal aluminium sample was provided by MTS System Corp., USA. The single crystal copper was grown using the Bridgman method and the nanocrystalline copper with average grain size 70 nm was prepared using electrodeposition technique (see [6] for details). Three BMG samples, $Ti_{40}Zr_{25}Ni_3Cu_{12}Be_{20}$ (BMG-1), $Ti_{41.5}Zr_{2.5}Hf_5Cu_{37.5}Ni_{7.5}Si_1Sn_5$ (BMG-2)and $Fe_{41}Co_7Cr_{15}Mo_{14}C_{15}B_6Y_2$ (BMG-3)were used in this study and the preparation details have been reported earlier [7-9].

All testing samples were mechanically polished to achieve a fine finishing before the nanoindentation experiments were performed at room temperature using a MTS nanoindenter XP system. Cyclic nanoindentation tests under constant peak load mode were performed at constant loading and unloading rate. The load-time sequence is shown in Figure 1(a). The load was held at its peak value for 1s for each cycle and the minimum load at the end of each cycle was set at 10 % of the preceding maximum load. The thermal drift rate, which was estimated from the 100s holding after unloading by 90%, was typically smaller than 0.05nm/s.

RESULTS AND DISCUSSION

Figure 1b shows a typical load-displacement curve obtained from the testing of a BMG-1 sample at the constant loading/unloading rate of 0.25mN/s for 100 cycles. The increasing displacement suggests an apparent time dependent displacement during the cycling indentation.

In order to investigate the reverse plasticity of materials under cyclic indentation, the first unloading-reloading cycle were excerpted and magnified in Figure 2 for all testing materials studied. Albeit a slight difference in the inclination, the first unloading-reloading curves obtained from the single crystal copper and nanocrystalline copper both show an obvious hysteresis loop, i.e. the unloading curve is steeper than the re-loading curve. It is also noted that the loop obtained from the nanocrystalline copper is slightly larger in displacement and has an earlier cross-over during reloading, after which the slope of the reloading curve gradually becomes smaller. This suggests that the plastic deformation during reloading occurs earlier in nanocrystalline copper than that in single crystal copper. A similar loop can been seen in Figure 2b for single crystal aluminium sample, which is less steep and with a even lower cross-over during reloading ("C" in Fig 2b) than that obtained from the nanocrystalline copper ("B" in Fig 2a).

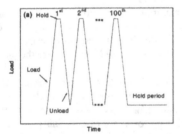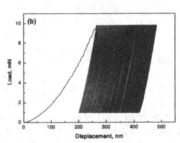

Figure 1. A typical load-time scheme for multi-cycle indentation (a) and a representative load-displacement curve obtained for a BMG-1 sample at the constant loading/unloading rate of 0.25 mN/s for 100 cycles (b).

It has been generally assumed that the nanoindentation unloading process involves the elastic recovery of the testing material [10]. However, experimental studies by Chowdhury and Laugier [11] have shown that the contact area and residual area can have significant difference, indicating non-elastic recovery. Likewise, the reloading is assumed to be an elastic deformation of the testing material and the plastic deformation resumes when the applied load reaches the preceding peak load. This would predict an overlap of the unloading and reloading curves. However, this is not the case. The unloading curves in Fig. 2a and 2b are steeper than the corresponding reloading curves. This is consistent with that reported by Shuman et al [2] that the elastic modulus measured from the unloading curve is higher than that measured from the reloading curve for different metals. The non-overlapping of the unloading and reloading curves suggests that there must involve some additional plastic deformation, i.e. reverse plasticity. However, it is still not clear whether the plastic deformation has been involved in only the unloading process or the reloading process, or both. Although Shuman et al [2] suggested the occurrence of the reverse plasticity during unloading, the deviation of the reloading curve from its linearity in the present study also suggests the association with plastic deformation during the reloading. The deviation and consequently the cross-over seem more obvious in single crystal aluminium than in nanocrystalline copper but are less obvious in single crystal copper. This may suggest that the resume of plastic deformation in a sequence of ease being single crystal aluminium, nanocrystalline copper and single crystal copper.

The unloading-reloading curves obtained from BMGs (Fig. 2c) differ from those from crystalline samples (Fig. 2a&b). As for the BMG-3, the unloading and reloading curves are found to be virtually indistinguishable in the present resolution, indicating the involvement of the same deformation level. The BMG-2 gives a small loop and the reloading curve is a little smaller than the unloading curve. BMG-1 gives a larger difference, particularly the upper part of the cycle. It is plausible that only elastic deformation has been involved in the unloading-reloading cycle in BMG-3 and a small amount of plastic deformation involves in BMG-2 and more in BMG-1. It is intriguing to notice that this sequence is the same as that of the glass transition temperature T_g for the three BMGs (838 K for BMG-3 [9]; 693 K for BMG-2 [8] and 605 K for BMG-1 [7]). In the same vein, it might be predicted that BMGs with T_g higher than 838K shall behave more elastically in the unloading and reloading cycles. Yang and Nieh [12] have discussed the physical analogy between the plastic deformation and glass transition of glassy materials and suggested that the increase of the glass transition temperature will result in the stronger glassy alloy, leading to larger energy required to overcome the bonding force between atoms and thus more elastic behaviour be expected.

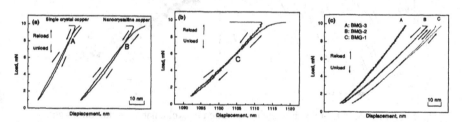

Figure 2 The unloading-reloading paths for the first cycle obtained in cyclic nanoindentation of single crystal copper and nanocrystalline copper (a), aluminium single crystal (b) and BMG-1, BMG-2, and BMG-3 (c).

To understand the effect of the cycling on the mechanical response of material, the individual unloading-reloading curves obtained from the BMG-1 sample after different number of cycles were compared and are shown in Figure 3a. To describe more closely the change, the maximum horizontal displacement disparity, δh, between the unloading curve and reloading curve has been measured and is shown in Figure 3b as a function of indentation time. The disparity δh decreases linearly with the indentation time, therefore the number of cycles.

It is known that the quenched-in free volume plays a very important role in plastic deformation of glassy alloys. The free volume evolution process during deformation under a shear stress has been described in Spaepen's free volume model as [13]

$$\dot{\upsilon} = \upsilon^* f \exp[-\frac{\Delta G^m}{kT}] \exp[-\frac{\alpha_s \upsilon^*}{\upsilon_f}] \times [\frac{2\alpha_s kT}{\upsilon_f S}(\cosh\frac{\tau\Omega}{2kT}-1)-\frac{1}{n_D}] \qquad (1)$$

where υ_f is the free volume at a particular site, Ω is the atomic volume, υ^* is the critical free volume required for an atomic jump, f is the jump frequency, ΔG^m is the activation energy for an atomic jump, τ is the applied shear stress, k is Boltzmann's constant, T is the temperature, α_f is a geometrical factor on the order of 1, and n_D is the number of atomic jumps required to annihilate the free volume υ^*. In Eq.(1), the accumulation and annihilation of free volume are characterized by the terms of $\frac{2\alpha_s kT}{\upsilon_f S}(\cosh\frac{\tau\Omega}{2kT}-1)$ and $(-\frac{1}{n_D})$, respectively. Free volume evolution is a

thermally activated process related to the temperature. In recent studies, it has been demonstrated that a great amount of heat can be generated by shear bands [14]. The local heating induced in turn facilitates the annihilation of free volume and results in microstructure transformation. Nanoindentation on a Zr-based metallic glass has shown to introduce crystallization both along the faces of the indent and in the region beneath the indent tip [15]. It is plausible that as the cyclic indentation proceeds, the accumulation of dissipated heat accelerates the free volume

relaxation and finally allows the balance between $\frac{2\alpha_s kT}{\upsilon_f S}(\cosh\frac{\tau\Omega}{2kT}-1)$ and $(-\frac{1}{n_D})$,

consequently minimizes the plastic deformation involved. The further study of cyclic deformation behaviour beyond 100 cycles will be an interesting subject of future work.

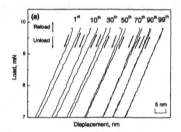 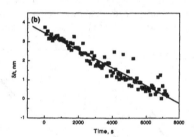

Figure 3 The unloading-reloading cycles obtained in cyclic nanoindentation (a) and corresponding relationship between δh and indentation time for BMG-1

To explore the loading rate effect, a series of cyclic tests were performed at different loading/unloading rates. Figure 4 presents the first unloading-reloading cycle excerpted from the cyclic indentation on the BMG-1 sample at the loading/unloading rate of 2.5, 1, 0.5 and 0.25mN/s. At high loading/unloading rates, the unloading and reloading curves form hysteresis loops and the loop width increases with increasing loading rate. This loop enlargement with increasing loading/unloading rate is still not clear. It is also noted that unloading and the reloading curves are roughly parallel to each other at high loading/unload rates, e.g. 2.5mN/s and 1mN/s. At low loading rates, the unloading and reloading curves form an open jaw shape and both curves become clearly deviated from each other at the lowest loading/unloading rate of 0.25mN/s. This suggests that faster loading/unloading rate may eliminate the reverse plasticity, resulting in parallel curves. Plastic deformation in BMGs usually occurs in the form of highly localized shear banding [13]. The loading rate dependence of the unloading-reloading cycle observed in the present work might be caused by a time dependent rearrangement of free volume, required by shear banding.

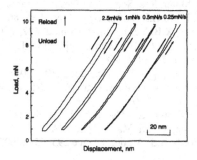

Figure 4 The first unloading-reloading paths obtained in cyclic nanoindentation for BMG-1 under different applied loading/unloading rates

CONCLUSIONS

The following conclusions can be drawn from the present study:
(1) The unloading-reloading cycles are different for different crystalline materials. The deviation between the unloading and reloading curves is the largest for aluminium single crystal, then nanocrystalline copper and the smallest for copper single crystal. In other words, the smaller the elastic modulus gives the larger deviation between the unloading and reloading curves, though the significance of the elastic modulus is not clear.
(2) The unloading-reloading cycles are different for different amorphous materials. Metallic glass with high T_g shows smaller deviation between the unloading and reloading curves.
(3) The deviation between the unloading and reloading curves gradually diminishes with the increasing number of cycles for BMG-1.
(4) Smaller loading/unloading rate leads to a larger deviation between the unloading and reloading curves for the BMG-1.

ACKNOWLEDGMENTS

The work described in this paper was partially supported by a grant from the University of Auckland, New Zealand (Project No. 9217/3609144) and a Program for New Century Excellent Talents in University, China.

REFERENCES

1. W.C. Oliver and G.M. Pharr, J. Mater. Res. 7, 1564 (1992).
2. D. J. Shuman, A. L. M. Costa, M. S. Andrade, Mater. Charac. 58, 380 (2007).
3. T.D. Raju, K. Nakasa, M. Kato, Acta Mater. 51, 457 (2003).
4. D. Pan, T. G. Nieh, M. W. Chen, Appl. Phys. Lett. 88, 161922 (2006).
5. B.X. Xu, Z.F. Yue, J. Wang, Mech. Mater. 39, 1066 (2007).
6. N.I. Tjahyono, Y.L. Chiu, Mater. Res. Soc. Symp. Proc. submitted (2007).
7. F.Q. Guo, H.J. Wang, S.J. Poon, G.J. Shiflet, Appl. Phys. Lett. 86, 091907 (2005).
8. Y.J. Huang, J. Shen, J.F. Sun, X.B. Yu, J. Alloys. Compd. 427, 171 (2007).
9. J. Shen, Q.J. Chen, J.F. Sun, H.B. Fan, G. Wang, Appl. Phys. Lett. 86, 151907 (2005).
10. Z.H. Xu, X.D Li, Acta Mater. 53, 1913 (2005).
11. S. Chowdhury, M. T. Laugier, Appl. Surf. Sci. 233, 219 (2004).
12. B. Yang, C. T. Liu, T. G. Nieh, Appl. Phys. Lett. 88, 221911 (2006).
13. F. Spaepen, Acta Metall. 25, 407 (1976).
14. C T Liu, L Heatherly, D S Easton, C A Carmichael, J.H. Schneibel, C.H. Chen, J.L. Wright, M.H. Yoo, J.A. Horton, A. Inoue, Metall. Mater. Tran. 29A, 1811 (1998).
15. J.-J. Kim, Y. Choi, S. Suresh, A.S. Argon, Science 295, 654 (2002).

Mater. Res. Soc. Symp. Proc. Vol. 1049 © 2008 Materials Research Society   1049-AA08-05

Uniaxial Compression Behavior of Bulk Nano-twinned Gold from Molecular Dynamics Simulation

Chuang Deng, and Frederic Sansoz
School of Engineering, University of Vermont, Burlington, VT, 05482

ABSTRACT

Parallel molecular dynamics simulations were used to study the influence of pre-existing growth twin boundaries on the slip activity of bulk gold under uniaxial compression. The simulations were performed on a 3D, fully periodic simulation box at 300 K with a constant strain rate of 4×10^7 s^{-1}. Different twin boundary interspacings from 2 nm to 16 nm were investigated. The strength of bulk nano-twinned gold was found to increase as the twin interspacing was decreased. However, strengthening effects related to the twin size were less significant in bulk gold than in gold nanopillars. The atomic analysis of deformation modes at the twin boundary/slip intersection suggested that the mechanisms of interfacial plasticity in nano-twinned gold were different between bulk and nanopillar geometries.

INTRODUCTION

The uniaxial deformation of micro/nano-sized gold pillars by nanoindentation has provided new fundamental insights into the influence of sample size on plasticity and strengthening in metals [1-3]. Clearly, it is important to understand the underlying mechanisms of deformation in gold at limited length scale. Molecular dynamics (MD) simulation has been commonly used to investigate the atomic-level mechanisms of plastic deformation at high strain rate in gold nanobeams and nanowires [4-10]. In recent MD work, Afanasyev and Sansoz [10] have studied the compression behavior of gold nanopillars (12 nm in diameter) consisting of nanoscale twin boundaries, a special type of grain boundary existing in this material. It was found that the presence of twin boundaries strongly influences the strength of gold nanopillars. Strengthening effects by interfacial plasticity were also found to largely depend on the twin interspacing. However, meaningful results related to the strength of gold nanopillars deformed by nanoindentation can only be accomplished if the influences of microstructure and sample size are fully understood. Particularly, further research effort must be undertaken to describe the bulk behavior of nano-twinned gold under uniaxial compression.

In this work, we used MD simulations to investigate the atomic mechanisms of plasticity in bulk nano-twinned gold under compression. A 3D, fully periodic simulation box was used to represent the bulk behavior of gold, excluding the effects of free surfaces. The deformation of single crystal Au was also considered. The next section describes the simulation method. The last section presents the effects of twin boundary on the compression behavior, and the atomic mechanisms operative at the intersection between slip and twin boundaries in bulk gold.

SIMULATION METHOD

Classical MD simulation was performed using LAMMPS Molecular Dynamics Simulator [11]. The interatomic potential used was an embedded-atom-method potential for FCC gold developed by Grochola et al. [12]. This potential was found to better predict the stacking fault and surface energies obtained experimentally, as opposed to other potentials in literature. The geometry of the model was a 200Å-wide cube with periodic boundaries in all three directions. The geometry and crystal orientations are shown in Figure 1. A typical model included about 0.5 million atoms. The simulations were performed in constant NVT ensemble at 300 K using a Nose/Hoover thermostat. The time step and entire simulation run were 5 fs and 2.5 ns, respectively. The model was first relaxed for 3000 steps (15 ps) to zero pressures along all three axial directions. Compression was performed on the relaxed structure by uniaxially shrinking the box at constant engineering strain rate (4×10^7 s^{-1}) while keeping the volume constant by proportionally expanding the box along the lateral directions. The loading direction was the [111] direction, normal to the plane of the twins (Figure 1). The compression stress was averaged both spatially over the total volume and temporally for 1000 steps (5 ps). The yield strength was calculated from the equivalent von-Mises stress at yielding, which takes into account the stress triaxiality along the three directions of space [13]. We simulated the deformation of 1 single crystal and 8 twinned structures whose twin interspacing varied from 2 nm to 16 nm.

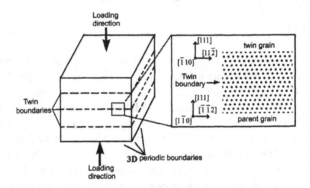

Figure 1. Atomic representation of bulk gold with three pre-existing growth twin boundaries. Close up view on the atomic structure and orientation at a twin boundary.

RESULTS AND DISCUSSION

Bulk compression behavior of single crystal and nano-twinned gold

Figure 2a represents the simulated stress-strain curves for single crystal and nano-twinned gold under uniaxial compression. This figure shows that the peak stress value corresponding to the limit of elasticity increased as the twin interspacing was decreased. It should be noticed that the yield strength values of all bulk samples (> 8 GPa) were found to be significantly higher than the strength of 12 nm-diameter gold nanopillars (< 6 GPa) reported elsewhere [10]. This difference can be attributed to the fact that the bulk behavior was simulated from a defect-free model while, in nanopillars, free surfaces play an important role as sources of dislocations, thus decreasing the yield stress. Furthermore, Figure 2b represents the variation of yield strength as a function of the twin interspacing. This figure shows a small, non-monotonic increase in yield strength as the twin interspacing is decreased. The yield strength values are equal to 8.15 GPa and 8.5 GPa for a twin interspacing of 16 nm and 2 nm, respectively, which corresponds to a 0.35 GPa increase in yield stress. By comparison, the increase in yield strength was found to be twofold larger in 12 nm-diameter gold nanopillars for the same twin interspacings [10]. It is also worth noting in Figure 2 that the yield strength of the single crystal model is equal to 8.21 GPa. Therefore, the yield strength did not drastically change between bulk single crystal and twinned gold. We can therefore conclude from our simulations that strengthening effects related to the twin size are less significant in bulk gold than in 12 nm-diameter gold nanopillars. This conclusion also supports the idea that the underlying mechanisms of interfacial plasticity at twin boundaries may be influenced by the sample size, resulting in differences between bulk and nanopillar geometries, as shown below.

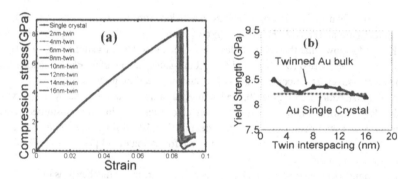

Figure 2. Bulk compression behavior of single crystal and nano-twinned gold by MD simulation. (a) Stress-strain curves. (b) Yield strength vs. twin interspacing in bulk gold.

Atomic mechanisms at the intersection between dislocation and twin boundary

Figure 3 shows the atomic structures of bulk gold with 8 nm-interspaced twin boundaries before and after a compression strain of ~ 9%. Intersection of slip dislocations with the twin boundaries was found to account for the differences in yielding phenomena. More specifically, in Fig. 3b corresponding to the plastically-deformed structure, it can be seen that a large number of stacking faults from the crystal lattice (highlighted by dark blue-colored atoms) exists at the intersection with twin boundaries. In this figure, it is also important to note the formation of <111> atomic steps on the twin planes. Such steps clearly show the occurrence of interfacial plasticity at the twin boundaries in simulated bulk gold.

(a) ε=0   (b) ε=0.0894

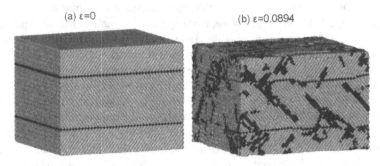

Figure 3. Atomistic model of bulk gold with 8 nm-interspaced twin boundaries under uniaxial compression. (a) Initial geometry before compression. (b) Structure with stacking faults and dislocations after yielding. Atoms are colored based on their centrosymmetry. Atoms in dark blue color represent stacking faults.

Atomistic details of the intersection between dislocations and twin boundaries at the onset of yielding are presented in Figure 4. In this figure, the twin interspacing corresponds to 8 nm. Figure 4a shows that the onset of plasticity is related to the activation of a $(11\bar{1})(011)$ slip in the grain separated by two twin boundaries. At this stage, each perfect <011> dislocation is dissociated into two <112> Shockley partial dislocations (a leading partial dislocation and a trailing partial dislocation). In Figure 4b, the leading partial is found to interact with the twin boundaries. It is shown in Figure 4c that the partial dislocations are directly transmitted across the upper and lower twin boundaries to form perfect (001)<110> dislocations in the twin grains. During this process, atomic steps corresponding to the formation of 1/3 [111] sessile Frank dislocations are left on the twin boundaries.

The atomic mechanism described above is different from the mechanisms of interfacial plasticity found in gold nanopillars under compression [10]. In the latter case, the mechanisms were related to the absorption and desorption [14] of both leading and trailing partials and the formation of Lomer-Cottrell locks and glissile twin dislocations that moved in the (111) plane of the twin boundary. A major difference is that glissile twin dislocations were not observed in the present study on bulk gold. Instead, the twin-slip reactions resulted in the formation of sessile Frank dislocations, which remained fixed on the twin boundary. This mechanism is in good agreement with experimental observations on bulk nano-twinned copper [15]. However, caution should be exercised here in comparing the different deformation modes at twin boundaries between bulk and nanopillar geometries, because the stress triaxiality in the bulk model may also play a role on the mechanisms.

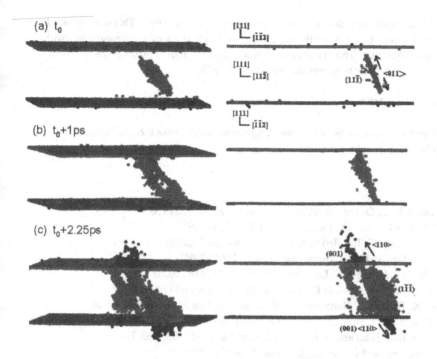

Figure 4. Atomistic details of the intersection between slip dislocations and twin boundaries in bulk nano-twinned gold. Atoms are colored based on their position with respect to the twin planes. Atoms in light blue and gold are present in the initial twin boundaries. Green-colored atoms between the two twin boundaries represent the slip dislocation. Atoms in red and blue colors belong to the transmitted dislocations. t_0 represents the time at the onset of yielding.

CONCLUSIONS

MD simulations have been carried out to characterize the slip behavior and plastic deformation of bulk nano-twinned gold at 300 K under uniaxial compression. We found in bulk nano-twinned gold simulated with 3D periodic boundary condition, that the limit of elasticity increased as the twin interspacing was decreased. However, strengthening effects related to the twin size were found to be less significant in bulk gold than in 12 nm-diameter gold nanopillars [10]. Our conclusions also suggested that the underlying mechanisms of interfacial plasticity at the intersection of dislocations with twin boundaries were different between bulk and nanopillar geometries. In the latter case, the reaction mechanisms at twin/slip intersection were related to the formation of Lomer-Cottrell locks and glissile twin dislocations that moved in the (111) plane of the twin boundary while, on bulk gold, the twin-slip reactions resulted in the formation of sessile Frank dislocations, which remained fixed on the twin boundary. This study showed that it is critically important to consider the sample size in the plasticity of metal nanopillars with nanoscale growth twins. This computational work may also help us interpret the results of nanoindentation experiments on compressed nanosized pillars.

ACKNOWLEDGEMENTS

Support from the Vermont Advanced Computing Center under Phase II NASA grant # NNG 06GE87G is gratefully acknowledged.

REFERENCES

1. J. R. Greer, W. C. Oliver and W. D. Nix, *Acta Mater.* **53**, 1821 (2005).
2. J. R. Greer and W. D. Nix, *Phys. Rev. B* **73**, 245410 (2006).
3. C. A. Volkert and E. T. Lilleodden, *Phil. Mag.* **86**, 5567 (2006).
4. J. Diao, K. Gall and M. L. Dunn, *Nano Lett.* **4**, 1863 (2004).
5. J.-S. Lin, S.-P. Ju and W.-J. Lee, *Phys. Rev. B.* **72**, 085448 (2005).
6. B. Hyde, H. D. Espinosa and D. Farkas, *JOM* September, 62 (2005).
7. J. Diao, K. Gall, M. L. Dunn and J. A. Zimmerman, *Acta Mater.* **54**, 643 (2006).
8. E. Rabkin and D. J. Srolovitz, *Nano Lett.* **7**, 101 (2007).
9. E. Rabkin, H.-S. Nam and D. J. Srolovitz, *Acta Mater.* **55**, 2085 (2007).
10. K. A. Afanasyev and F. Sansoz, *Nano Lett.* **7**, 2056 (2007).
11. S. J. Plimpton, *J. Comp. Phys.* **117**, 1 (1995); http://lammps.sandia.gov/
12. G. Grochola, S. P. Russo and I. K. Snook, *J. Chem. Phys.* **123**, 204719 (2005).
13. W. F. Hosford. *Mechanical Behavior of Materials*, Cambridge University Press (2005)
14. T. Zhu, J. Li, A. Samanta, H. G. Kim and S. Suresh, *Proc. Natl. Acad. Sci. U. S. A.* **104**, 3031 (2007).
15. L. Lu, Y. Shen, X. Chen, L. Qian and K. Lu, *Science* **304**, 422 (2004).

Nanomechanics of Polymers,
Time Dependent Characterization

Mater. Res. Soc. Symp. Proc. Vol. 1049 © 2008 Materials Research Society 1049-AA09-05

Interfacial Characterization of Multiple Layer Coatings on Thermoplastic Olefins (TPO)

Aaron M Forster[1], Chris A. Michaels[2], Justin Lucas[1], and Lipiin Sung[1]

[1]Materials and Construction Research Division, National Institute of Standards and Technology, 100 Bureau Dr., Gaithersburg, MD, 20899

[2]Surface and Microanalysis Science Division, National Institute of Standards and Technology, 100 Bureau Dr., Gaithersburg, MD, 20899

ABSTRACT

Thermoplastic olefins (TPO) have made significant inroads as polymeric materials for interior and exterior automotive parts. Spray applied chlorinated polyolefins (CPO) are often used to improve paint adhesion to the low surface energy TPO substrates. The penetration of the CPO into the substrate is difficult to quantify, but is critical to achieving a good paint/TPO bond. The interphase between each layer in a coated TPO coupon was investigated using a combination of instrumented indentation and confocal Raman microscopy. The degree of CPO interpenetration was altered by using three different CPOs exposed to two different manufacturing methods for the same TPO and base/clear coat system. It was found that the interfaces, CPO/base coat or CPO/TPO, were chemically and mechanically sharp at the 1 μm lateral resolution of both techniques. A gradient in the modulus through the thickness of the clear coat was observed using instrumented indentation.

INTRODUCTION

Thermoplastic olefin is a general term used within the automotive industry to describe a polypropylene blended with a phase separated ethylene-propylene or ethylene-butene rubber. TPOs are used as substrates for interior and exterior parts. Polypropylene is a brittle material that has low impact strength. The dispersed rubber phase improves the toughness of polypropylene by halting the propagation of crazes within the matrix. The advantages of TPO compared to steel include lower cost, extended durability, recycling of used parts, and the ability to form parts into complicated shapes and textures. Despite the multiple advantages of TPO, there are two main difficulties: TPO is sensitive to processing conditions and its low surface energy makes it difficult for automotive coatings to bond to it.

The prevalent view of the TPO micro-structure is given in Figure 1a [1-3]. The bulk TPO is composed of a dispersed rubber phase, with rubber particles having diameters between 1 μm and 20 μm. The rubber phase morphology shifts from spherical to elliptical or fibril in a 2 μm to 3 μm region below the surface. At the near surface is a polypropylene rich layer with a shear induced crystalline structure. This is a consequence of shear and rapid cooling during the injection molding process. Ryntz [4] has shown that processing variables such as gate volume, injection speed, and mold temperatures can have a significant impact on the thickness and morphology of this near surface region.

If a strong adhesive bond is not created at the base coat/TPO interface, then water, solvents, detergent, and impact damage may initiate coating failure and decrease service life. A common method used to improve adhesion is to spray apply low molar mass chlorinated polyolefin (CPO) primers. The solvent in the CPO solution penetrates the polypropylene and permits the diffusion and entanglement of the CPO within the dispersed rubber phases [1,5,6]. Mirabella et al. [7] utilized thermodynamic calculations to predict a CPO penetration into the TPO of 11 nm at 25° C and 5000 nm at 120 ° C. Subsequent, scanning transmission X-ray microscopy of CPO coated TPO samples showed an interfacial thickness of 350 nm [7]. Several additional studies have found the thickness of the interfacial region to range between 500 nm and 20 μm [1-4, 8,9].

Instrumented indentation has been shown as an effective tool to measure the mechanical properties through the thickness of TPO coated materials [10]. In this work, we further investigate the thickness and properties of the interphase between the TPO, CPO and top coat layers using a combination of instrumented indentation (IIT) and confocal Raman. Confocal Raman microscopy is a non-destructive optical technique that allows generation of surface compositional maps using the unique spectroscopic signatures of the various chemical constituents. This technique yields information comparable to that acquired from secondary ion mass spectrometry (SIMS) yet is generally not subject to sample damage problems (e.g. ion induced cross-linking) that can complicate SIMS analysis of polymeric materials. High lateral resolution can be achieved with this technique (<1 μm) and little sample preparation is required. The confocal Raman and instrumented indentation have similar length scale resolution across each interface for these materials. Thus, the combination of techniques is intended to quantify the impact of interpenetration on chemical and mechanical properties near the interface for industrially relevant materials and processing conditions. This has been difficult to achieve in the past for these materials.

EXPERIMENTAL[†]

In this work, CPOs were provided by Eastman Chemical. The TPO panel coupons were assembled and painted at Visteon. The CPO primer increases adhesion with top coat paints and a graphic of the laminate structure is shown in Figure 1. The dry layer thicknesses, as reported by Visteon, were base coat (35 μm), clear coat (17 μm), and CPO (7.5 μm to 10 μm). Two sets of samples were made: unmounted coupons and epoxy mounted cross-sectioned samples. The cross-sections were polished using a mechanical grinding process with silicon carbide grinding papers (successive grits of 320, 500, 800, 1200, 4000) and a final polish using a l μm suspension polish.

Three different CPO materials were investigated. The control CPO, designated CNTL, contained ≈ 22 mass % chlorine. The high chlorine CPO, designated Cl, contained 3 mass % more chlorine, but the same molecular weight as the CNTL. The high molecular weight CPO, designated MW, contained the same molar mass of chlorine as CNTL, but a higher molecular weight. The base/clear coat materials were polyester/acrylic blends cured using an umpigmented melamine crosslinker. Two different processing conditions were investigated: bake and wet-on-wet. Typically, TPO parts are painted using a wet-on-wet process [1]. The CPO, base coat, and clear coat were spray applied and the solvents allowed to flash evaporate between applications for a period of 3 minutes, 5 minutes, and 10 minutes, respectively. The coated part is baked at

---

[†] Certain instruments or materials are identified in this paper in order to adequately specify experimental details. In no case does it imply endorsement by NIST or imply that it is necessarily the best product for the experimental procedure.

120 °C for 30 minutes. In the bake processing, the CPO was baked at 120 °C for 30 minutes after application. The base coat and clear coat application, flash time, final bake time and temperature were the same as the wet-on-wet application.

Instrumented Indentation

Measurements were performed using a commercial nanoindenter (MTS Nanoinstruments, NanoXP). In order to achieve a higher resolution placement of the indents, the sample was mounted with the laminate at an angle to the Y-axis that ranged between 10° and 15°, see Figure 1b. Indents were conducted in a vertical line (Y-axis motor only) progressing from the epoxy towards the TPO. The spacing between indents was 15 μm, which translated to between 2.5 μm and 4 μm lateral movement (X-direction) toward the TPO interface. Three lines of 60 indents (180 indents/sample) were used to create modulus and hardness maps. The absolute position of each indent and its relationship to each interface was measured using Laser Scanning Confocal Microscopy (LSCM) [10]. The position resolution is 0.37 μm/pixel at 50X. The interface between the CPO and base coat was used as the zero reference position because it was easily defined. Drzal et al. [10] used a similar procedure to investigate the modulus of coated TPO panels via instrumented indentation.

A 1 μm radius, 60° diamond cone indenter was used to indent the samples. Loading was performed at a constant strain rate of 0.05 s^{-1}. The stiffness of the tip-sample contact was continuously measured during indentation by imposing a small oscillation of 2 nm at a frequency of 45 Hz [11]. The reported values of modulus for each indent were averaged over a depth range from 500 nm to 1000 nm without a drift correction. The maximum contact radius was ≈ 1 μm and the measured modulus was constant within this depth range. The Poisson's ratio was assumed constant and equal to 0.35. Several methodologies are available to utilize load, displacement, and stiffness data to determine modulus and hardness [12-14].

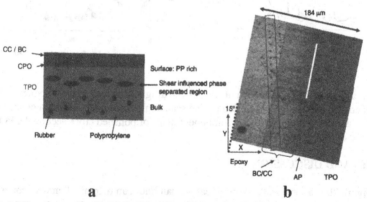

Figure 1: a) Microstructure of a TPO plaque and the laminate construction of a painted TPO. (not to scale) b) LSCM composite image (50X) showing a line of indents through the laminate sample. The white line is a guide for the CPO/base coat interface.

Confocal Raman Microscopy

Details of the confocal Raman microscope used to evaluate the cross-sectioned samples are described elsewhere [15]. Figure 2a shows the Raman spectra for each layer (epoxy, base/clear coat, CPO, TPO) in the mounted laminate. The methylene torsion band of polypropylene (PP) at 1220 cm^{-1} was chosen for mapping the TPO layer [16], as this relatively weak band shows little overlap with the CPO spectrum. The presence of talc complicates the mapping results for the TPO layer. Regions of high talc concentration (talc can be mapped using the 366 cm^{-1} or 679 cm^{-1} band [17]) yielded smaller levels of PP, leading to large apparent concentration fluctuations in the maps of the TPO region shown below. The variations seen in the mapping on the TPO region is largely a reflection of a change in the local concentration of talc. The C-Cl stretching bands in the 660 cm^{-1} to 730 cm^{-1} region were chosen as the marker band for the CPO adhesion promoter. This spectral feature is the main band that distinguishes the CPO from TPO and thus is the only realistic choice for this species. The band area in this case was calculated by integrating over the high frequency half of the peak and doubling this area. This was done to avoid interference with a strong Si-O-Si talc band at 679 cm^{-1}. The base coat and clear coat layers were indistinguishable as the pigment was removed from the basecoat to avoid compromising the index matching between the fluid and the sample. The carbonyl band at 1730 cm^{-1} was chosen as the marker band for these two indistinguishable layers. Finally the 1116 cm^{-1} band, presumably an oxirane wag vibration, is used to map the mounting epoxy.

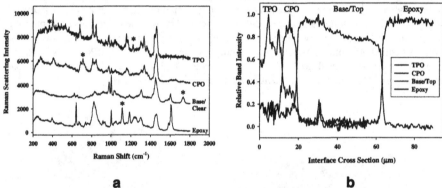

a   **b**

Figure 2: a) 785 nm Raman spectra of the pure coating system materials. Chemically specific marker bands (*) used in the generation of line scan component maps. Spectra offset vertically for visual clarity. b) Raman line scan component maps of polished cross-sectional CPO Mw, bake sample of the coating interfaces.

RESULTS AND DISCUSSION

Figure 2b is a representative confocal Raman line scan map of a laminate containing the higher molecular weight CPO prepared using the bake method. The band area values at each point in the line scan are normalized to the maximum value for that band to enable comparison between species. At x = 63 μm, the interface between clear coat and mounting epoxy is sharp.

The small peak in the TPO and CPO profiles at x = 31 μm is due to a change in the slope of the fluorescence background in this region which leads to a non-zero band intensity. This peak is purely an artifact of how this strongly shifted background influences the TPO and CPO band area calculations. The origin of the changing fluorescence background is unknown but might well be due to some small surface contaminant. The slow decline in band intensity across the base/clear coat region is likely due to a small drift of the sample with respect to the laser focus (data acquisition across this region took approximately 4 hours), slightly reducing the overlap between the laser and the sample.

In the range between 0 μm and 10 μm the predominant species is TPO although both the CPO and base/clear coat bands show a non-zero value. Careful inspection of the spectra in this region reveals that this non-zero value is due to minor spectral overlap with the TPO marker band and is not due to the presence of the base coat or CPO in the TPO region. The large variation in the TPO band area in this range is largely due to variation in the talc concentration and reflective of the heterogeneity of TPO as opposed to measurement noise. The sharp TPO-CPO interface can be seen clearly at about x = 12 μm where the CPO value quickly increases to near max value and the TPO band area drops to a lower level. The TPO value in the CPO region is again due to the minor presence of spectral intensity in the CPO spectrum in the TPO marker band region. This is not surprising given the chemical similarity between the CPO and TPO. These complications aside, the important observation is that this interface is quite sharp with a measured width of less than 1 μm, a value that is certainly limited by the instrument resolution. There is also no evidence of large scale diffusion of CPO into the TPO layer at the micrometer length scale, although it is important to note that diffusion at levels below a few percent would likely fall below the sensitivity of this measurement. Similarly, sharp interfaces were found for CPO/TPO interface in the CNTL sample.

Figure 3 shows line scans of indentation for the three different coatings prepared with the bake processing of the CPO. The CPO interfaces are sharp. Similar to the TPO region measured with Raman spectroscopy, the modulus within the TPO varies significantly. The current protocol does not appear sensitive enough to measure a modulus gradient in the TPO as a result of processing.

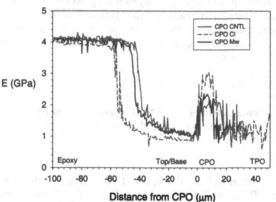

Figure 3: Line scan maps of modulus through the thickness of the laminate sample. Sample prepared with the bake method and all three CPOs listed. Zero is the base coat /CPO interface.

Indentation was able to distinguish all three CPO layers from the adjacent layers and confirm the CPO interfaces for both the bake and wet-on-wet samples (not shown) were sharp. Measurement resolution is at the micrometer length scale. An advantage of instrumented indentation over traditional hardness and mechanical property evaluation techniques is observed in the base/clear coat layer in Figure 3. The modulus of the clear coat increases as the indenter moves toward the epoxy/clear coat interface. At the interface, the epoxy can confine the movement of the clear coat underneath the indenter and increase the contact stiffness. A general estimate of the size of the stress field around the sphere for axisymmetric indentation is 2 times the contact radius [18, 19]. For this test geometry, indentations within 4 μm of an interface are within this zone. As seen in Figure 3, the modulus begins to increase $\approx$ 15 μm from the epoxy interface. The surface hardening observed here is believed to originate from skin formation induced from solvent evaporation during the spray process [20]. This information is valuable for determining the processing steps to increase the surface modulus of coatings [21].

It is possible that polishing damages the coatings by smearing interfaces or inducing surface hardening. This was not believed to have a significant impact on the measurements for two reasons. First, the modulus was constant within the indent depth tested. Second, the perpendicular surface of the sample, unmounted in epoxy, was indented to a depth of 2 μm using a 10 μm 90° conical tip. The modulus was measured in the depth range of 1 μm to 2 μm and named *top down*. The top down modulus was (1.66 $\pm$ 0.013) GPa for the CNTL CPO compared to (1.63 $\pm$ 0.017) GPa for the cross-section sample. In addition, the top down modulus for the Cl CPO was (1.55 $\pm$ 0.022) GPa and (1.69 $\pm$ 0.25) GPa for the cross-section. The MW CPO top down modulus was (1.57 $\pm$ 0.012) GPa and (1.76 $\pm$ 0.034) GPa for the cross-section. Note that the modulus of the cross-sectioned sample was taken 4 μm from the epoxy/film interface. The qualitative agreement supports the conclusion that the gradient in modulus is not related to sample geometry.

CONCLUSIONS

The interface between the different layers within a painted TPO panel was chemically and mechanically characterized as a function of CPO type and manufacture process. It was found that, within the resolution of both instruments, the CPO interface was sharp and independent of CPO type. Instrumented indentation showed an increase in the modulus of the clear coat layer near the surface.

ACKNOWLEDGEMENTS

This work was conducted through the NIST/Industry Polymer Interface Consortium. The authors thank Dr. Peter Drzal for development of initial measurements on these systems.

REFERENCES

1. H. R. Morris, B. Munroe, R. A. Ryntz, P. J. Treado, *Langmuir* 14 2426 (1998).
2. H. R. Morris, J. F. Turner II, B. Munro, R. A. Ryntz, P. J. Treado, *Langmuir* 15 2961 (1999).
3. Y. Ma, M. A. Winnik, P. V. Yaneff, R. A. Ryntz, *JCT Research* 2 407 (2005).
4. R. A. Ryntz, *J. of Vinyl & Additive Technology* 3 295 (1997).

5. R. A. Ryntz, *JCT Research* **3** 3 (2006).
6. D. J. Burnett, F. Thielmann, R. A. Ryntz, *JCT Research* **4** 211 (2007).
7. F. M. Mirabella, N. Dioh, C. G. Zimba, *Polymer Engineering and Science* **40** 2000 (2000).
8. H. Tang, D. C. Martin, *J. of Mat. Sci.* **37** 4783 (2002).
9. J. E. Lawniczak, K. A. Williams, L. T. Germinario, *JCT Research* **2** 399 (2005).
10. P.L. Drzal, L-P Sung, D. Brintz, R. A. Ryntz, in *International Coatings for Plastics Symposium Proceedings* June 6-8, (2005).
11. A. C. Fischer-Cripps, *Nanoindentation*, Springer-Verlag, New York, 2004
12. I. N. Sneddon, *Int. J. Engng Sci.* **3** 47 (1965).
13. W. C. Oliver, G. M. Pharr, *J. Mater. Res.* **7** 1564 (1992).
14. M. R. VanLandingham, *Journal of Research of the National Institute of Standards and Technology* **108** 249 (2003).
15. X. Gu, C.A. Michaels, D. Nguyen, Y.C. Jean, J.W. Martin and T. Nguyen, *Appl. Surf. Sci.* **252** 5168 (2006).
16. H. Tadokoro, M. Kobayashi, M. Ukita, K. Yasufuku, S. Murahashi, and T. Torii, *J. Chem. Phys.* **42** 1432 (1965).
17. G.J. Rosasco and J.J. Blaha, *Appl. Spec.* **34** 140 (1980).
18. Johnson, K. L., *Contact Mechanics* Cambridge University Press: Cambridge, (1985).
19. C. Y. Hui, A. Jagota, Y. Y. Lin, E. J. Kramer, *Langmuir 18 1394 (*2002).
20. M. Duskova-Smrckova, K. Dusek, *J. Mat. Sci.* **37** 4733 (2002).
21. B. Flosbach, *Macromol. Symp.* **187** 503 (2002).

Mater. Res. Soc. Symp. Proc. Vol. 1049 © 2008 Materials Research Society 1049-AA07-08-OO08-08

Viscoelastic Behavior of a Centrally Loaded Circular Film Being Clamped at the Circumference

Michelle L Oyen[1], Kuo-kang Liu[2], and Kai-tak Wan[3]

[1]Engineering Department, Cambridge University, Trumpington St, Cambridge, CB2 1PZ, United Kingdom

[2]ISTM School of Medicine, Keele University, Stoke-on-Trent, ST4 7QB UK, United Kingdom

[3]Mechanical and Industrial Eng, Northeastern University, 360 Huntington Ave, Boston, MA, 02115

ABSTRACT

A new theoretical model is constructed for the viscoelastic response of a clamped circular membrane deformed by a spherical indenter, using the classical Maxwell and Standard Linear Solid (SLS) constitutive equations. Preliminary stress-relaxation experiments are performed for a hydrogel membrane and the corresponding data fitted to the SLS-based viscoelastic model.

INTRODUCTION

Mechanical characterization of biological tissues, membranes and cells is an important tool to understand many phenomena in life-sciences and to promote biomedical engineering. For instances, hardening of the glycoprotein shell of a mouse egg due to fertilization, stiffening of prosthetic cornea constructs as a result of cell growth, lack of interfacial adhesion at epithelial cells and the basement lamina membrane leading to malignant metastasis, cell aggregation leading to tissue formation etc. Conventional indentation method is limited to *hard* materials and coating, and is not suitable for compliant polymer films, ultra-thin biomembranes, and membranous cells and organelles etc. Besides, the localization damaged zone around the indenter-substrate interface does not capture the global behavior and viscoelasticity of, for example, a single cell with an encapsulating membrane of thickness in the order of 10-15nm. A variant form of indentation for freestanding membrane is devised to circumvent the limitations.

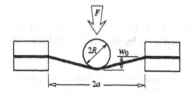

Figure 1. Sketch of a shaft indentation for quantifying elastic and viscoelastic behavior of the sample membrane.

Figure 1 shows a thin delicate membrane clamped by two identical rings to form a diaphragm. An external load is applied to the film center giving rise to a conical deformed geometry. Quasi-static and cyclic loading by the conventional or micro-force universal testing machines allows one to measure the viscoelastic response. Creep test over a prolonged period of time can also be conducted by dropping a ball bearing of specific radius and weight to the sample center. The materials parameters are then found by relating the measured applied load, shaft displacement and deformed profile to the constitutive relation. By choosing the convenient dimensions and ratio of the shaft radius and diaphragm radius / thickness, the multi-scale mechanical behavior can be mapped. The smallest possible sample can be made in micron scale if the clamping rings are fabricated by standard photolithography.

THEORY

The constitutive relation is first derived using linear elasticity, then by linear viscoelasticity. A thin film with bending rigidity, $\kappa = Eh^3 / 12(1-v^2)$, elastic modulus, E, and Poisson's ratio, v, thickness, h, radius, a, and a tensile pre-stress, σ_0, is clamped at the periphery. An external force, P, applied to the film center via a shaft with radius, R ($<< a$) leads to a central displacement, δ, and a concomitant membrane stress, σ. An exact analysis shows

$$\delta = \frac{P}{2\pi(\sigma+\sigma_0)h}\left\{\frac{1-\beta\ I_0(\beta)-\beta\ K_1(\beta)+\beta^2 I_0(\beta)K_1(\beta)}{\beta\ I_1(\beta)}+K_0(\beta)-\log\left(\frac{2}{\beta}\right)+\gamma\right\} \quad (1)$$

where $\beta = [(\sigma + \sigma_0)\,h\,a^2 / \kappa]^{1/2}$, I_n and K_n are the nth order modified Bessel functions of the first and second kinds, and $\gamma = 0.577216$ is the Euler-Mascheroni constant [1]. For thin and flexible membranes with virtually zero bending moment and $\sigma_0 = 0$, (1) is reduced to

$$P = \frac{\pi}{4}\left(\frac{E}{1-v}\right)\left(\frac{h}{a^2}\right)\delta^3 \quad (2)$$

Note that (2) is valid only when $\delta > 2h$, or, $\lambda > 350$ with $\lambda = 24.3^{1/2}.\ (1-v^2)^{3/2}\ P\,a^2 / Eh^4$ [3]. For $\lambda < 350$, plate bending dominates and the conical profile no longer applies since the profile gradient at the circular clamp must vanish.

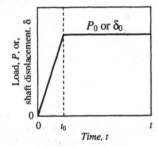

Figure 2. Ramp and hold input for fixed-rate loading followed by hold.

Before deriving the full solution, a few basic guidelines are stated for conversion of the linear elastic governing equations to viscoelastic [4]. The general solution for an isotropic material incorporates a pair of time-independent parameters, such as $E(t)$ and $v(t)$, or, shear modulus, $G(t)$, and bulk modulus, $K(t)$. Experimental investigations can be done in the common creep test with constant load, relaxation test with constant displacement, and loading at various load- or displacement rates. Analysis of the conventional indentation is nicely summarized by Johnson [5] and the practical implementation by Oyen [6-9] in a series of recent works. In an

158

elastic problem, the summed normal components of stress and strain are related by K such that σ_{ii} = $3\,K\,\varepsilon_{ii}$, and the deviatoric stress and strain are related by G such that $s_{ij} = 2\,G\,d_{ij}$. Hereafter, G = $E/3$ is taken for incompressible membranes (v = 0.5). Contrary to linear elasticity, viscoelasticity is strongly tied to the mode of control (i.e. displacement- or load-control). The hereditary integrals are needed to relate $G(t)$ to the shear creep compliance, $J(t)$, by

$$ d_{ij}(t) = \frac{1}{2} \int_0^t J(t-u)\frac{ds_{ij}(u)}{du}\,du \qquad \text{and} \qquad s_{ij}(t) = 2 \int_0^t G(t-u)\frac{dd_{ij}(u)}{du}\,du \qquad (3) $$

for load-control and for displacement-control testing, respectively. Note that $\hat{G}(s)\hat{J}(s) = 1/s^2$ in Laplace space. For sufficiently fast loading, i.e. time to reach peak displacement, δ_0, or peak load, P_0, approximates to one tenth the material time constant, the applied loading becomes

$$ P(t) = P_0\,H(t) \qquad \text{or} \qquad \delta(t) = \delta_0\,H(t) \qquad (4) $$

where the Heaviside step function is defined as $H(t<0) = 0$ and $H(t\geq 0) = 1$. For a sample bearing the linear elastic constitutive relation in the form of $P = k\,E\,\delta^n$ with k a function accounting for the geometry, and for finite and constant loading rates r as $P_0 = r\,t_0$ and v as $\delta_0 = v\,t_0$, full integrals must be evaluated:

$$ P(t) = 6k \int_0^t G(t-u)\frac{d\delta^3(u)}{du}\,du \qquad \text{or} \qquad \delta^3(t) = \frac{1}{6k} \int_0^t J(t-u)\frac{dP(u)}{du}\,du \qquad (5) $$

Figure 2 shows the loading ramp. Closed-form solutions are obtained for Maxwell fluid (M) and Standard Linear Solid (SLS), based on their constitutive relations (Table 1).

Table 1: Basic relationships for Maxwell and standard linear solid viscoelastic models.

	Maxwell	Standard Linear Solid
Constitutive equation	$\eta\dot{\sigma} + G\sigma = G\eta\dot{\varepsilon}$	$\sigma(G_1 + G_2) + \dot{\sigma}\eta = G_1 G_2 \varepsilon + G_1\eta\dot{\varepsilon}$
Relaxation function, $G(t)$	$G(t) = G\exp\left(\dfrac{-t}{\tau_M}\right)$, $\tau_M = \dfrac{\eta}{G}$	$G(t) = \dfrac{G_1}{G_1 + G_2}\left[G_2 + G_1\exp\left(\dfrac{-t}{\tau_{RS}}\right)\right]$, $\tau_{RS} = \dfrac{\eta}{G_1 + G_2}$
Creep function, $J(t)$	$J(t) = \dfrac{1}{G} + \dfrac{t}{\eta}$	$J(t) = \dfrac{1}{G_1} + \dfrac{1}{G_2}\left[1 - \exp\left(\dfrac{-t}{\tau_{CS}}\right)\right]$, $\tau_{CS} = \dfrac{\eta}{G_2}$

For shaft indentation of an incompressible membrane, (2) is rewritten as

$$P = k E \delta^3 \tag{6}$$

where $k(h,a)$ is defined to be $k = (\pi/2).(h/a^2)$ with $v = 0.50$ for most incompressible polymeric materials. Note that the indenter radius R can be incorporated into k, though the mathematically involved function is not given here. The major difference between the current analysis and the classical Hertz contact is the $P \sim \delta^3$ dependence in the former instead of $P \sim \delta^{3/2}$ in the latter. In case of instance loading,

$$P(t) = 6 k\, G(t)\, \delta_0^3 \qquad \text{or} \qquad \delta^3(t) = \frac{P_0}{6k} J(t) \tag{7}$$

In case of ramping load, $\delta(t) = v\, t$ or $P(t) = r\, t$,

$$\frac{dP(u)}{du} = r \qquad \text{or} \qquad \frac{d\delta^3(u)}{du} = 3v^3 u^2 \tag{8}$$

Here a two phase hereditary integral condition would be evaluated as:

$$\delta^3(t) = \frac{1}{6k} \int_0^{t_0} J(t-u)\frac{dP(u)}{du} du \;+\; \frac{1}{6k} \int_{t_0}^{t} J(t-u)\frac{dP(u)}{du} du \tag{9}$$

The first integral from 0 to t_0 is evaluated but the second integral is zero since $d\delta^3/du = dP/du = 0$ for $P_0 = r\, t_0$ or $\delta_0 = v\, t_0$ for $t \geq t_0$. Table 2 shows the results of integration.

Table 2: Mechanical response for membrane indentation

	Ramp ($0 \leq t \leq t_0$)	Hold ($t \geq t_0$)
Maxwell, creep	$\delta^3(t) = \dfrac{r}{6k}\left[\dfrac{t}{G} + \dfrac{t^2}{2\eta}\right]$	$\delta^3(t) = \dfrac{r}{6k}\left[\dfrac{t_0}{G} + \dfrac{t_0 t}{\eta} - \dfrac{t_0^2}{2\eta}\right]$
Maxwell, relaxation ($\tau \equiv \tau_M$)	$P(t) = 18kGv^3 \{t^2\tau - 2t\tau^2 + 2\tau^3[1-\exp(-t/\tau)]\}$	$P(t) = 18kGv^3 \exp^{-t/\tau} *$ $\left(\exp^{t_0/\tau}\left[t_0^2\tau - 2t_0\tau^2 + 2\tau^3\right] - 2\tau^3\right)$
SLS, creep ($\tau \equiv \tau_{CS}$)	$\delta^3(t) = 6kr\,[C_2 t - C_3\tau(1-\exp^{-t/\tau})]$	$\delta^3(t) = 6kr\left[\begin{array}{l} C_2 t_0 - \\ C_3\tau\exp(-t/\tau)(\exp(t_0/\tau)-1) \end{array}\right]$
SLS, relaxation ($\tau \equiv \tau_{RS}$)	$P(t) = 6kC_0 v^3 t^3 + 18kC_1 r^3 \left[\begin{array}{l} t^2\tau - 2t\tau^2 \\ + 2\tau^3\left(1-\exp^{-t/\tau}\right) \end{array}\right]$	$P(t) = 6kC_0 v^3 t_0^3 + 18kC_1 v^3 \exp^{-t/\tau} *$ $\left(\exp^{t_0/\tau}\left[t_0^2\tau - 2t_0\tau^2 + 2\tau^3\right] - 2\tau^3\right)$

EXPERIMENT

A 2% solution of sodium alginate was formed by dissolving 2 g of Protanal LF200 S (FMC BioPolymer, Norway) into 100 ml of deionised water. The ratio of β-D-mannuronic (M block) to α-L-guluronic (G block) in this type of alginate was attuned to be 0.23 [10]. A membrane of the type of aliginate hydrogel which has thickness 0.66 mm was carefully clapmed by two annulus plates covered with two rubber O-rings, which are used to avoid unwanted stresses generated during clamping. A glass-made spherical indenter of 600 μm which was driven by a micro-stepping motor (ESP300, Newport, Irvine, CA) and then moved at a "ramp" speed (500 μm/sec) to apply a loading force at the center of the clapmed membrane. The central displacement was then fixed at 1000 μm and the force of the indentation was measured by a force transducer (404A, Aurora Scientific Inc, Canada) linked to the indenter upto 45 minutes. The experimntal measurement of the stress relaxtion data fitted with a theortical curve is shown as Figure 3. For simplicity, only the curve described by using stand linear solid model is shown in the Figure. The tedious parameter identification of the variables shown in Table 2 will be discussed elsewhere.

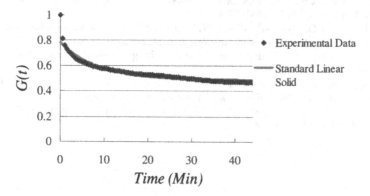

Figure 3. Experimental stress-relaxation data of an alginate hydrogel fitted standard linear solid model

CONCLUSION

A new viscoelastic model is developed for the shaft indentation of a 2-D axisymmetric membrane clamped at the periphery according to the Maxwell and Standard Linear Solid models. Experimental investigation of a circular hydrogel membrane shows the expected viscoelatic behaviors. The technique and models are particularly useful for thin delicate polymer films where the conventional tensile methods are not suitable.

ACKNOWLEDGEMENTS

KTW acknowledged the support by US National Science Foundation (CMMI # 0757140). KKL would like to thank Dr. M. Ahearne for help in the experimental work.

REFERENCES

1. Wan KT, Guo S, Dillard DA, "A theoretical and numerical study of a thin clamped circular film under an external load in the presence of tensile residual stress", *Thin Solid Films* **425**, 150-162 (2003).
2. Wan KT, Mai YW, "Fracture mechanics of a shaft-loaded blister of thin flexible membrane on rigid substrate", *International Journal of Fracture* **74**, 181-197 (1995).
3. Liu KK, Wan KT, "Multi-scale mechanical characterization of a freestanding polymer film using indentation", submitted (2007).
4. Mase GT and Mase GE: *Continuum Mechanics for Engineers*, 2nd Ed., CRC, Boca Raton, FL, 1999.
5. Johnson KL, *Contact Mechanics*. Cambridge University Press, UK, 1985.
6. Oyen ML, "Analytical Techniques for Indentation of Viscoelastic Materials", *Philosophical Magazine* **86** [33-35], 5625 – 5641, 2006.
7. Mattice JM, Lau AG, Oyen ML, Kent RW, "Spherical Indentation Load-Relaxation of Soft Biological Tissues", *Journal of Materials Research* **21**, 2003-10 (2006).
8. Oyen ML, "Spherical Indentation Creep Following Ramp Loading", *Journal of Materials Research* **20**, 2094-2100 (2005).
9. Oyen ML, Bushby AJ, Viscoelastic Effects in Small-Scale Indentation of Biological Materials. *International Journal of Surface Science and Engineering* **1**, 180-197 (2007).
10. Drury JL, Dennis RG, Mooney D J, "The tensile properties of alginate hydrogels", *Biomaterials* **25**, 3187-3199 (2004).

Mater. Res. Soc. Symp. Proc. Vol. 1049 © 2008 Materials Research Society   1049-AA10-02

Analysis of Indentation Creep

D. S. Stone[1], and A. A. Elmustafa[2,3]

[1]Materials Science and Engineering, University of Wisconsin-Madison, 1509 University Avenue, Madison, WI, 53706

[2]Department of Mechanical Engineering, Old Dominion University, Norfolk, VA, 23529

[3]The Applied Research Center, Old Dominion University, Newport News, VA, 23606

ABSTRACT

Increasingly, indentation creep experiments are being used to characterize rate-sensitive deformation in specimens that, due to small size or high hardness, are difficult to characterize by more conventional methods like uniaxial loading. In the present work we use finite element analysis to simulate indentation creep in a collection of materials whose properties vary across a wide range of hardness, strain rate sensitivities, and work hardening exponents. Our studies reveal that the commonly held assumption that the strain rate sensitivity of the hardness equals that of the flow stress is violated except for materials with low hardness/modulus ratios like soft metals. Another commonly held assumption is that the area of the indent increases with the square of depth during constant load creep. This latter assumption is used in an analysis where the experimenter estimates the increase in indent area (decrease in hardness) during creep based on the change in depth. This assumption is also strongly violated. Fortunately, both violations are easily explained by noting that the "constants" of proportionality relating 1) hardness to flow stress and 2) area to (depth)^2 are actually functions of the hardness/modulus ratio. Based upon knowledge of these functions it is possible to accurately calculate 1) the strain rate sensitivity of the flow stress from a measurement of the strain rate sensitivity of the hardness and 2) the power law exponent relating area to depth during constant load creep.

INTRODUCTION

Indentation tests have long been used to study rate sensitive deformation in solids. The great majority of work has been performed at high homologous temperatures where solids tend to be soft, that is, where the flow, or yield, stress is a small fraction of the Young's modulus. Because of the work on materials under conditions where they are soft, there have been a number of theoretical treatments that address the issue of rate sensitive hardness measurements under conditions where the elastic deformations are small and therefore relatively unimportant [1-4].

The present research is motivated by the desire to investigate rate sensitive deformation in hard materials like refractory coatings and bulk metallic glasses where the yield stress and hardness constitute a significant proportion of the Young's modulus. Our investigation differs from earlier studies because in the hard materials which we now consider the elastic deformations become large and therefore important. For low yield stress materials the indentation creep properties are not sensitive to modulus, so the modulus can therefore be neglected. For hard materials the modulus effects must be taken into account.

In an indentation creep test the hardness, $H(t)$, is defined as instantaneous load divided by instantaneous projected contact area, $L(t)/A(t)$. The effective strain rate during indentation creep can be calculated a number of different ways but we use $\dot{\varepsilon}_{eff} = (1/\sqrt{A})d\sqrt{A}/dt$. The strain rate sensitivity of the hardness is then

$$v_H = \frac{\partial \ln H}{\partial \ln \dot{\varepsilon}_{eff}}\bigg|_{x_p}. \tag{1}$$

where the derivative is taken at fixed temperature and depth [5]. The results of the analysis for strain rate sensitivity do not depend appreciably which characteristic length of the indent (e.g., depth, square root of area) is used to define strain rate [6].

The purpose of the present paper is to analyze indentation creep in terms of the evolution of indent area and hardness during an experiment. Finite element analysis is used to simulate the indentation creep test. Load and depth-vs.-time data obtained from the finite element analysis simulations can then be analyzed the same way that real experimental data are, except that with finite element analysis we have the advantage of knowing what the projected contact area is at any instant in time, whereas in a real experiment the area must be estimated from load-depth data or inferred from dynamic stiffness measurements [7].

Here we review the results of the analysis for creep. We find that the trend of v_H/v_σ as a function of $H/E*$ is a direct consequence of the relationship between the hardness and the yield stress. Usually, the hardness is said to be about 3 times the yield stress, and this is a good approximation for soft materials. More correctly, however, the ratio $k = H/\sigma_Y$ is a function of $\sigma_Y/E*$ (or $H/E*$), so that as the stress below the indenter changes during a creep experiment, so does the ratio k. We shall also examine the growth of the area of an indent during indentation creep. It is often assumed in the analysis of indentation creep that the square root of the area increases in direct proportion to the total depth (e.g., [8, 9]). It turns out that this assumption is valid during loading but breaks down during constant load creep; and any analysis that employs the assumption can lead to erroneous conclusions. Fortunately, however, the growth of the area as a function of depth is only sensitive to $H/E*$, and its trend with $H/E*$ can be easily understood using a simple analysis; that analysis allows the experimenter to identify the correct relationship between area and depth during creep.

FINITE ELEMENT ANALYSIS SIMULATIONS

Finite element analysis was used to perform indentation creep simulations [6]. The simulations employed the elastoplastic feature of ABAQUS finite element code. In the finite element model, the indenter and solid were modeled as bodies of revolution to take advantage of the axisymmetry of conical indentation. The indenter is considered a perfectly rigid cone with an inclined face angle of $22.5°$. The sample is modeled as an axisymmetric elastic-plastic von Mises material. Quadrilateral axisymmetric 4 node isoparametric elements were used to model the semi-infinite medium. Meshes were generated using GENMESH2D and multi-point constraints (MPCs) were created using GENMPC2D. The mesh was further refined in regions near the edge of contact to minimize the effect caused by the discrete jump during relaxation.

The contact between the indenter and the sample was modeled as having a friction coefficient, μ, varying from 0.0 to 0.5 [10]. To simulate a variety of material behaviors we adopted a strain hardening creep law given by

$$\dot{\varepsilon}^{cr} = (\kappa\sigma^n[(m+1)\varepsilon^{cr}]^m)^{\frac{1}{m+1}}, \qquad (2)$$

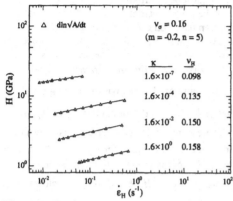

Figure 1) Hardness-strain rate data generated from computer simulations. The units of κ are $GPa^{-5}s^{-0.8}$.

where n, m are the stress and the strain exponents. κ is a rate parameter: by adjusting κ while keeping n and m constant, it is possible tailor the hardness of the material while maintaining fixed the strain rate sensitivity $\nu_\sigma = (m+1)/n$ and work hardening exponent $\chi = -m/n$.

The indentation process was simulated in four sequential steps. In the first step, contact between indenter tip and sample was established. In the second step the indenter was pushed into the sample to a depth of approximately 0.3 μm under a load that increased parabolically with time. In the third step, the load on the tip was held constant for 2-3 seconds (indentation creep). The last step consisted of pulling the indenter out of the sample.

Hardness vs. strain rate curves for materials all with the same strain rate sensitivity ν_σ (same m, n) but with different hardness (different κ) are shown in Figure 1. In this figure the slopes of the curves, which correspond to ν_H, are all different. We may therefore conclude that in general $\nu_H \neq \nu_\sigma$. Nevertheless, the slopes approach ν_σ as the hardness decreases, and we find that the ratio of ν_H / ν_σ is a unique function of H / E^* independent of the other material properties as shown in Figure 2. The simulations in Figure 2 are for a frictionless indenter-specimen contact; the addition of friction, however, does not appreciably affect the results [10]. The curve in this figure is an empirical fit of the data given by

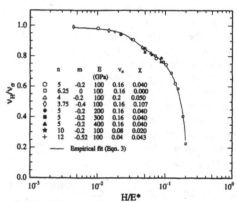

Figure 2. Effect of hardness level on strain rate sensitivity of the hardness.

$$v_H / v_\sigma \cong \left(0.91 - 0.057 \arctan(\frac{\log(x/28.2)}{0.2}) \right) \times \sqrt{1 - (x/0.21)^2} , \tag{3}$$

from which it is possible to correct a measurement of v_H to obtain v_σ.

In the finite element simulations it is possible to calculate instantaneous hardness based on load and area as direct outputs of the simulations. In a real experiment, however, it is only possible to measure load and depth directly, and area must be inferred from an indirect measurement. Dynamic stiffness measurements can be used to determine area [7]; but usually experimenters estimate the change of area during creep based on depth of penetration by assuming that the area is proportional to the square of depth [8, 9]. This latter practice is open to dispute. For instance, Rar and coworkers [11] have reported that the area-(depth)^2 law is violated during creep, an assertion supported by our work. We define the parameters ζ_i $= \partial \ln \sqrt{A} / \partial \ln h_i \big|_L$, where h_i is the measure of depth being used to estimate area. In general, it is possible to rely on total measured depth, h_t, plastic depth, h_p [12], or contact depth, h_c [13]. A plot of the ζ_i determined from simulations of different materials at zero friction is shown in Figure 3. Interestingly, all of the data fall along master curves, but only for low $H/E*$ do all of the ζ_i approach 1. For $H/E* < 0.04$ ζ_p is closest to 1 (approximately 1.03; our previous measurements have relied on ζ_p [5]). For hardness higher than $H/E* \approx 0.04$ ζ_c is closest to 1.

DISCUSSION

Our results show that in general, the strain rate sensitivity of the hardness differs from that of the flow stress. The results also show that the assumption that square root of area increases in proportion to depth during constant load creep is strongly violated at higher hardness. Fortunately, to aid in the analysis of indentation creep the trends in the data in Figure 2 and 3 are both systematic and easily understood. The trend in v_H / v_σ, Figure 2, is predominately a result of the proportionality factor between hardness and flow stress (k in $H = k\sigma$), which

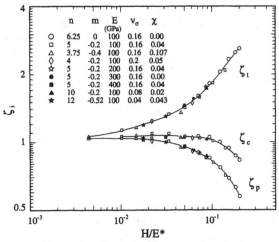

Figure 3. Exponents relating $\sqrt{A}$ to depth (h_t, h_p, h_c) during indentation creep.

varies as a function of the ratio of $\sigma / E*$ where σ is the flow stress in the plastic zone beneath the indenter. Changes in hardness during indentation creep derive not only from changes in σ but also from changes in k. Thus, in terms of logarithms, the change in of hardness during indentation creep is given by

$$\Delta \ln H = (1 + \frac{d \ln k}{d \ln(\sigma / E^*)}) \Delta \ln \sigma. \qquad (4)$$

For soft (low H / E^*) materials the factor k is independent of σ / E^* in which case $\Delta \ln H = \Delta \ln \sigma$. In this instance $v_H = v_\sigma$. For elastic materials with H / E^* approaching 0.21, the hardness becomes proportional to E^* rather than σ, so that the second term in parentheses in Equation 4 approaches -1, in which case $\Delta \ln H$ (and therefore v_H) goes to zero. A more detailed analysis is given elsewhere [6]. In short, the trend in Figure 2 is simply because the material is becoming increasingly elastic at higher H / E^*.

Likewise, the trends in the different ζ_i, Figure 3, can also be explained by noting that the elasticity of the solid affects the ratios of $\sqrt{A} / h_i$. For instance, we may suppose $\sqrt{A} = C_i h_i$ where C_i depends on H / E^*. During creep, the change in $\sqrt{A}$ with H and h_i can be written

$$\Delta \ln \sqrt{A} = (\frac{d \ln C_i}{d \ln(H / E^*)} \Delta \ln H + \Delta \ln h_i). \qquad (5)$$

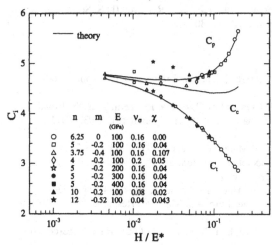

During constant load creep $d \ln H = -2d \ln \sqrt{A}$; algebraic manipulation of (5) leads to

$$\frac{d \ln C_i}{d \ln(H / E^*)} = \frac{1}{2}(\frac{1}{\zeta_i} - 1). \qquad (6)$$

To verify this relationship we show data of C_p, C_t and C_c in Figure 4. The lines in Figure 4 for C_i-vs.-H / E^* ("theory") have been generated by taking spline fits of the ζ_i-vs.-H / E^* data (lines in Figure 3), then incorporating them into Equation 6 and integrating. Our ability to generate the lines in Figure 4 from the lines in Figure 3 proves that values of ζ_i in Figure 3 come from the trends in C_i with hardness.

Figure 4. Proportionality factors relating $\sqrt{A}$ to depth (h_t, h_p, h_c) from simulated hardness tests on different materials.

Interestingly, there is a significant amount of scatter in the C_i for H / E^* small in Figure 4. This scatter is due to the influence of pileup, which is sensitive to the work hardening and strain rate sensitivity properties of the materials. In comparison there is relatively little scatter in ζ_i for low H / E^*, which suggests that while pileup might affect the ratio $C_i = \sqrt{A} / h_i$, pileup has little effect on the derivative $d \ln \sqrt{A} / d \ln h_p$. We conclude that, during indentation creep at constant load, the deviations of the ζ_i away from 1 are caused by the changes taking place in C_p, C_t, and C_c as hardness is lowered. These changes can be predicted based on experiments performed on a collection of materials at different $H/E*$.

CONCLUSIONS

In conclusion, an analysis of indentation creep at constant load has been performed. The strain rate sensitivity of the hardness generally differs from that of the flow stress. The growth of the area of the indent with respect to depth differs from what it does during loading. Fortunately, with both situations the trends are insensitive to work hardening, creep, and friction, and the trends in behavior can be understood in terms of how the elastic-plastic problem depends upon the ratio $H/E*$. This knowledge helps in the measurement and interpretation of rate-sensitive deformation using nanoindentation.

ACKNOWLEDGMENTS

A.A. Elmustafa wishes to thank the Summer Faculty Fellowship Research Program Office of Research, Old Dominion University Grant # 993003. Research (D.S. Stone) sponsored by the National Science Foundation (Award CMS-0528073).

REFERENCES

1. Y.-T. Cheng and C.-M. Cheng. Materials Science & Engineering R: Reports, 2004. **R44**(4-5): p. 91.
2. A.F. Bower, N.A. Fleck, A. Needleman and N. Ogbonna. Proceedings of the Royal Society of London, Series A (Mathematical and Physical Sciences), 1993. **441**(1911): p. 97.
3. S.N.G. Chu and J.C.M. Li. Journal of Materials Science, 1977. **12**(11): p. 2200.
4. R. Hill. Proceedings of the Royal Society of London, Series A (Mathematical and Physical Sciences), 1992. **436**(1898): p. 617.
5. D.S. Stone and K.B. Yoder. Journal of Materials Research, 1994. **9**(10): p. 2524.
6. A.A. Elmustafa, S. Kose and D.S. Stone. Journal of Materials Research, 2007. **22**(4): p. 926.
7. D.L. Goldsby, A. Rar, G.M. Pharr and T.E. Tullis. Journal of Materials Research, 2004. **19**(1): p. 357.
8. D. Jang and M. Atzmon. Journal of Applied Physics, 2003. **93**(11): p. 9282.
9. F. Wang, P. Huang and K.W. Xu. Applied Physics Letters, 2007. **90**(16): p. 161921.
10. A.A. Elmustafa and D.S. Stone. Journal of Materials Research, 2007. **22**(10): p. 2912.
11. A. Rar, S. Sohn, W.C. Oliver, D.L. Goldsby, T.E. Tullis and G.M. Pharr. *On the measurement of creep by nanoindentation with continuous stiffness techniques.* (2005 Fall MRS SYmposium Proceedings, Vol 841, Boston, MA) p. 119-124.
12. M.F. Doerner and W.D. Nix. Journal of Materials Research, 1986. **1**(4): p. 601.
13. W.C. Oliver and G.M. Pharr. Journal of Materials Research, 1992. **7**(6): p. 1564.

Mater. Res. Soc. Symp. Proc. Vol. 1049 © 2008 Materials Research Society

Haasen Plot Activation Analysis of Constant-Force Indentation Creep in FCC Systems

Vineet Bhakhri, and Robert J. Klassen

Mechanical & Materials Engineering, The University of Western Ontario, London, N6A 5B9, Canada

ABSTRACT

An analysis of the length-scale and temperature dependence of Haasen plots obtained from constant-force nano- and micro-indentation creep tests are reported here. The operative deformation mechanism for all the systems studied involved dislocation glide limited by dislocation-dislocation interactions. Our findings illustrate the potential usefulness of Haasen plot activation analysis for interpreting data from constant-force pyramidal indentation creep tests.

INTRODUCTION

The use of Haasen plot activation analysis to characterize the operative plastic deformation mechanisms in materials subjected to uniaxial loading is well established [1-5]. This type of analysis involves plotting the experimentally determined inverse activation area $1/\Delta a$ of the deformation process versus the applied shear stress τ. The resulting "Haasen plot" gives information on the operative mechanism of plastic deformation. Linear trends of $1/\Delta a$ and τ indicates that dislocation glide in the test material is being limited by dislocation/dislocation interactions (i.e. the Cottrell-Stokes law is maintained) while nonlinear trends arise from the cumulative effect of several types of obstacles or to a changing microstructural state within the sample during the test.

The application of Haasen plot activation analysis to interpret indentation creep test data is very attractive from the view point of understanding local variations in plastic deformation across complex microstructures [6]. Indentation test data are, however difficult to interpret due to the complex stress and strain states within the indentation plastic zone.

During a Constant Force (CF) pyramidal indentation test the indenter applies a high local stress state, represented by the average indentation stress σ_{ind}, to the indented sample. The sample creeps and this causes the indentation depth h to increase. Since the indentation force is held constant σ_{ind} decreases as h increases. The average indentation stress σ_{ind} and local indentation strain rate $\dot{\varepsilon}_{ind}$ can be expressed, for a Berkovich indenter, as

$$\sigma_{ind} = \frac{F}{\alpha A(h)} \text{ and } \dot{\varepsilon}_{ind} = \frac{\dot{h}}{h} \tag{1}$$

where α is an empirically derived function accounting for the sink-in/pile-up effects around the indenter and A(h) is the area function of the indenter probe. Both σ_{ind} and $\dot{\varepsilon}_{ind}$ are average values representing the complex distribution of $\dot{\varepsilon}$ and σ within the indentation plastic zone. We can represent these distributions by an equivalent average indentation shear stress, τ_{ind} and an equivalent average indentation shear strain rate $\dot{\gamma}_{ind}$ as

$$\tau_{ind} = \frac{\sigma_{ind}}{3\sqrt{3}} \text{ and } \dot{\gamma}_{ind} = \sqrt{3}\dot{\varepsilon}_{ind} = \frac{\sqrt{3}\dot{h}}{h} \tag{2}$$

Since the local stress around the indentation is very large, the underlying deformation mechanism, in most ductile metals, is one involving obstacle-limited dislocation glide. The relationship between $\dot{\gamma}_{ind}$ and τ_{ind} for this deformation mechanism can be expressed as [7-10]

$$\dot{\gamma}_{ind} = \dot{\gamma}_0 e^{\frac{-\Delta G(\tau_{eff})}{kT}} \tag{3}$$

where $\dot{\gamma}_0 = 10^{-11}(\tau_{ind}/\mu)^2$ (s^{-1}) [11], $\tau_{eff} = \tau_{ind} - \tau_{threshold}$, k is the Boltzmann constant, T is the temperature and $\Delta G(\tau_{eff})$ is the thermal activation energy required for the dislocations to overcome a deformation-rate limiting obstacle. The energy that must be supplied to the dislocation to overcome an obstacle is $\Delta G_0 = \Delta G(\tau_{eff} = 0)$ and can be expressed as the sum of $\Delta G(\tau_{eff})$ and the work performed on the dislocation by τ_{eff} as

$$\Delta G_0 = \Delta G(\tau_{eff}) + \tau_{eff} b \Delta a \tag{4}$$

$\dot{\gamma}_{ind}$ and τ_{eff} data are obtained from a CF indentation creep test therefore Equation (3) can be used to determine $\Delta G(\tau_{eff})$, ΔG_0 and Equation (4) can then be used to calculate the apparent activation area Δa (b^2).

We present here the results of CF nano-indentation creep tests performed on annealed (100) Au single crystal and cold-worked polycrystalline Au to investigate the length-scale dependence of the shape of the $1/\Delta a$ versus τ_{eff} Haasen plots of Au in both the annealed and heavily cold-worked condition. We then reanalyse previously reported high temperature micro-indentation creep data to determine the dependence of the shape of the $1/\Delta a$ versus τ_{eff} Haasen plots upon temperature for three Al-based systems.

170

EXPERIMENT

The CF nanoindentation creep tests reported here were performed at 300 K and were one hour in duration. The Au surfaces were mechanically and chemically polished to ± 30 nm surface roughness. The (100) Au single crystal was then annealed at 1123 K in a H_2 reducing atmosphere to ensure that no residual plastic strain or surface contamination remained from the polishing stages. 20% cold-work was induced in compression in a polycrystalline Au specimen.

Between four and eight nanoindentation creep tests were performed at each of the initial indentation depths of $h_0 = 200$, 400, 800, and 2000 nm on both the annealed Au and the cold-worked Au samples using pyramidal berkovich indenter probe. The indenter was loaded at a certain rate such that it reaches the specified h_0 between 15-20 seconds, then the load was held constant at that depth for 3600 seconds followed by unloading at original rate. Indenter was held at a minimal load value for 200 seconds during unloading in every test to measure the thermal drift of the system which is accounted in the analysed data. The creep tests were performed with a Nano-Test indentation testing platform made by Micro Materials Ltd (Wrexham, UK).

RESULTS & DISCUSSION

Length-scale dependence of deformation kinetics

The σ_{ind} decreased continuously with time during the one-hour CF nanoindentation tests performed on the Au samples. σ_{ind} at the beginning of the creep tests was higher when h_0 was small. This is in agreement with the known length-scale dependence of the hardness of Au [12]. σ_{ind} at the end of the creep test was much smaller in magnitude than at the start of the test and less dependent upon indentation depth.

Figure 1 shows the Haasen plots of $b^2/\Delta a$ versus τ_{eff} calculated, using Equations (3 and 4) for the data from the annealed (100) Au single crystal and the 20% cold-worked polycrystalline Au. The large scatter in the data arises from the large indent-to-indent scatter in the $\dot{\varepsilon}_{ind}$ versus σ_{ind} data. This is presumably due to the sensitivity of nanoindentation test to local variations in roughness of the surface of the test sample. For both materials, however, the $b^2/\Delta a$ versus τ_{eff} data pass through the origin; moreover, the data from individual tests show linear trends of $b^2/\Delta a$ versus τ_{eff} however there is considerable scatter in the slopes of these trends.

Our observation that the Haasen plots from individual tests are linear and pass through the origin confirm that the operative creep deformation mechanism is controlled by dislocation-dislocation interactions in both the annealed and the cold-worked Au [1,2]. This finding is reasonable for an FCC metal such as Au used in this study. The fact that the range of slope of the data from individual creep tests is the same for deep and shallow indentations in annealed and cold-worked samples suggests that there is no significant length-scale or material state dependence, beyond the experimental scatter shown in Figure 1.

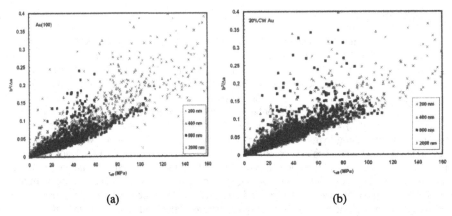

(a)   (b)

Figure 1: Haasen plots of; (a) annealed (100) Au and (b) 20% Cold-worked polycrystalline Au.

Temperature dependence of deformation kinetics

Haasen plot activation analysis was performed on previously reported data from CF microindentation tests performed at 473, 573, 673, 773, and 833 K on three Al-based systems: 1) commercially pure Al, 2) annealed wrought 2024 Al alloy, and 3) annealed Powder Metallurgy (PM) fabricated 2024 Al alloy [6, 7]. Figure 2 shows the resulting Haasen plots. The data in these plots indicate much less scatter than the data from the nanoindentation creep tests (Figure 1). This is due to deeper indentations used in these tests which were less affected by small surface roughness and therefore displayed less indent-to-indent scatter.

The $b^2/\Delta a$ versus τ_{eff} trends in Figure 2 show nonlinear trends which are temperature-dependent however all the trends pass through the origin. The curves, for a given temperature, are not clearly different in shape or magnitude amongst the three Al compositions tested. These observations lead us to conclude that the operative deformation mechanism is one involving obstacles that are equally present in all three materials; namely, dislocation-dislocation interactions. The tangent of the $b^2/\Delta a$ versus τ_{eff} curves is small early in the creep test when τ_{eff} is high but increases as τ_{eff} decreases with increasing time. The mechanical activation energy ($\Delta W = \tau_{eff} b \Delta a$) required to be applied to a dislocation to overcome an obstacle is inversely related to the slope of the Haasen plot [2]. The mechanical activation energy decreases with time and at a particular τ_{eff} value, increases with temperature during CF microindentation creep of Al and Al alloys. Increasing ΔW with temperature suggests that indentation creep rate controlling obstacles bear a long-range tail profile [2].

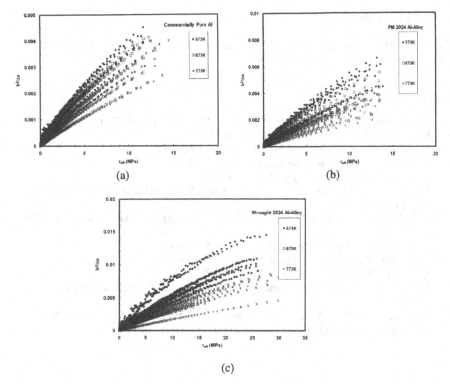

Figure2: Haasen plots derived from previously reported high temperature CF microindentation creep tests performed on; (a) commercially pure Al, (b) PM 2024 Al alloy and (c) Wrought 2024 Al alloy [9,10].

CONCLUSION

We have reported here an analysis of the length-scale and temperature dependence of $b^2/\Delta a$ versus τ_{eff} Haasen plots obtained from CF indentation creep tests performed on annealed and cold-worked Au and on three Al-based metals. The operative deformation mechanism for all the systems studied involved dislocation glide limited by dislocation-dislocation interactions. The change in mechanical activation work between microindentation tests performed at different temperatures could be seen from changes in the tangent of the respective Haasen plots.

These findings illustrate the potential usefulness of Haasen plot activation analysis for interpreting data from CF indentation creep tests.

ACKNOWLEDGMENTS

Authors wish to thank the Natural Engineering Science & Research Council of Canada for providing financial support for this research.

REFERENCES

1. U.F. Kocks, A.S Argon and M.F Ashby, Prog. Mater Sci. **19**, 1 (1975)
2. R.A. Mulford, Acta Metalurgica **27**, 1115 (1979)
3. H. Mecking, U.F. Kocks, Acta Metallurgica **29**, 1865 (1981)
4. S. Siamoto, H. Sang, Acta Metallurgica **31**, 1873 (1983)
5. W. Bochniak, Acta Metallurgica **41**, 3133 (1993)
6. M.F. Tambwe, D.S. Stone, A.J. Griffin, H. Kung, Y. Cheng Lu and M. Natasi, J. Mater. Res. **14**, 407 (1999)
7. W.B. Li, J.L. Henshall, R.M. Hooper and K.E. Easterling, Acta Metall. Mater. **39**, 3099 (1991)
8. V. Bhakhri, R.J. Klassen, Scripta Mater **31**, 395 (2006)
9. V. Bhakhri, R.J. Klassen, J. Mater. Sci. **41**, 2249 (2006)
10. V. Bhakhri, R.J. Klassen, J. Mater. Sci. **41**, 2259 (2006)
11. H.J. Frost, M.F. Ashby, "Deformation-Mechanism Maps" (Pergamon Press, Oxford, 1982), pp.21
12. Z. Zong, J. Lou, O.O. Adewoye, A.A. Elmustafa, F. Hammad and W.O. Soboyejo, Mater. Sci. Eng. A **434**, 178 (2006)

Mater. Res. Soc. Symp. Proc. Vol. 1049 © 2008 Materials Research Society   1049-AA05-17

Mechanical Properties of Sputtered Silicon Oxynitride Films by Nanoindentation

Yan Liu, I-Kuan Lin, and Xin Zhang
Department of Manufacturing Engineering, Boston University, 15 St. Mary's Street,
Brookline, MA, 02446

ABSTRACT

Silicon oxynitride (SiON) films with different oxygen and nitrogen content were deposited by RF magnetron sputtering. Fourier-transform infrared (FT-IR) spectroscopy study revealed that co-sputtered SiON films were composed of one homogeneous phase of random bonding O-Si-N network. Time-dependent plastic deformation (creep) of SiON films were investigated by depth-sensing nanoindentation at room temperature. Results from nanoindentation creep indicated that plastic flow was relatively less homogenous with increasing nitrogen content in film composition. A deformation mechanism based on atomic bonding structure and shear transformation zone (STZ) plasticity theory was proposed to interpret creep behaviors of sputtered SiON films.

INTRODUCTION

Silicon oxynitride (SiON) has been under extensive investigation as a promising material system for the development of optics, photonics and microelectronics applications. SiON demonstrates unique tunability in optical and electronic properties when changing the chemical composition of oxygen and nitrogen, from silicon dioxide (SiO_2) to silicon nitride (Si_3N_4). The flexibility in mechanical properties allows SiON films with low residual stress and high thermal stability to be integrated into the micro-electro-mechanical systems (MEMS) [1]. However, only a few attempts have been made to study the mechanical properties of SiON thin films. And even more, time-dependent plastic deformation (creep) behavior of SiON thin film at room temperature has not been reported previously in literature. Depth-sensing nanoindentation is a powerful technique for studying the mechanical properties of materials in micro- or nano-scaled dimensions. Upon precisely recording the load-displacement data and constructing relevant analysis models, nanoindentation has been used to reveal and interpret the mechanisms of various mechanical properties [2].

In the present work, SiON films with different composition of oxygen and nitrogen content were deposited by RF magnetron sputtering. Creep behaviors of sputtered SiON films have been investigated by depth-sensing nanoindentation with constant-load controlled experiments carried out at room temperature. Energy dispersive X-ray (EDX) spectroscopy and Fourier-transform infrared (FT-IR) spectroscopy were employed to characterize SiON films with respect to stoichiometric composition and atomic bonding structure. A deformation mechanism based on chemical bonding structure and shear transformation zone (STZ) amorphous plasticity theory was proposed to interpret nanoindentation creep properties of SiON films.

EXPERIMENT

Silicon oxynitride films with a thickness of approximately 1 μm were deposited on silicon wafer substrates at room temperature by a Discovery 18 RF magnetron sputtering system. Among five sputtered SiON specimens, silicon oxide and silicon nitride films were directly sputtered from pure (> 99.9% purity) silicon oxide and silicon nitride targets in the presence of argon plasma, respectively. Three SiON films with different nitrogen/oxygen ratios were deposited by co-sputtering the silicon oxide and silicon nitride targets by arbitrarily adjusting the RF powers (listed in Table I.) applied to the two targets. The flow rate of argon was 25 SCCM (standard cubic centimeters per minute at STP) to maintain a sputtering pressure at 2 mTorr during the deposition. In this study, SiO_x, SiN_y and SiO_xN_y were adopted to specifically represent the sputtered silicon oxide, silicon nitride, and co-sputtered silicon oxynitride films, respectively. All SiON films were subjected to a post-deposition thermal treatment at 400 °C for 30 minutes in a nitrogen ambient environment.

Table I. Sputtering RF power applied in SiON film deposition, and corresponding atomic concentration of Si, O and N in film composition.

Specimen	RF Power (W)		Atomic Concentration (at. %)			
	SiO_2	Si_3N_4	Si	O	N	N/(O+N)
SiO_x	300	0	37.34	62.66	0	0
SiO_xN_y1	300	100	51.23	43.17	5.60	11.48
SiO_xN_y2	300	300	41.39	32.36	26.25	44.79
SiO_xN_y3	100	300	42.44	22.24	35.32	61.36
SiN_y	0	300	46.04	(5.74)	48.23	100

Scanning electron microscope (SEM) equipped with an energy dispersive x-ray (EDX) spectrometer was employed to quantitatively analyze the compositional content of the sputtered SiON films. The chemical bonding and microstructure within the SiON films was identified by Fourier transform infrared spectroscopy (FT-IR). A K-Br light splitter and a DTGS detector were employed in the FT-IR analysis to obtain the mid-infrared (MIR) absorbance spectra (400 – 4000 cm^{-1}) of SiON films in a nitrogen ambient environment.

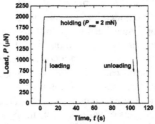

Figure 1. Load scheme of the nanoindentation creep experiment for the SiON films.

Nanoindentation creep experiments of sputtered SiON films were conducted on a Triboindenter nanomechanical test system using a constant-load controlled mode at room temperature. A standard Berkovich tip with a tip radius of approximately 150 nm and a half-angle of 65.35° was chosen for the experiments. The scheme of constant-load controlled indentation creep is demonstrated in Figure 1, in which the maximum indent load was 2 mN, the holding period was 100 s, and both constant loading rate and unloading rate were 400 μN/s. Thermal drift rate was monitored before each individual indentation was initiated, and compensation was automatically applied to the indentation results by the Triboindenter.

DISCUSSION

EDX and FT-IR analysis of sputtered SiON films

The corresponding element concentrations of Si, O and N in atomic percentage (at.%) obtained by EDX analysis are summarized in Table I, indicating that all of the SiON films exhibited a Si-sufficient non-stoichiometric composition, which could be a result of oxygen and/or nitrogen byproducts formed in the sputtering process. The small amount of oxygen (5.74 at.%) detected in the SiN_y film could be attributed to the absorption of free oxygen in the surface of the film, evidenced by the absence of absorbance peaks related to Si-O-Si bond in FT-IR spectrum of SiN_y film (Figure 2).

For non-stoichiometric amorphous SiO_xN_y films, two models known as the random mixture model (RMM) and the random bonding model (RBM) have been proposed to depict the chemical bonding structures [3]. The SiO_xN_y film with RMM structure is comprised of a random mixture of SiO_2 and Si_3N_4 separate phases on a microscopic scale. In the SiO_xN_y film with RBM structure, Si atoms are randomly bonded with O and/or N atoms to form a homogenous O-Si-N network with five different tetrahedral coordinations, in which $y = 4 - x$, and x may vary from 0 to 4 [3]

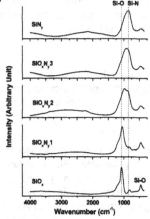

Figure 2. MIR FT-IR absorbance spectra of the SiON films.

Figure 2 presents the MIR FT-IR absorbance spectra of sputtered SiON films. Three characteristic features of Si-O-Si bonding group can be identified in the FT-IR spectrum of SiO_x film [4]. The strongest absorbance peak (with a shoulder) at approximately 1075 cm⁻¹ was the Si-O stretching mode. The weakest absorbance peak near 810 cm⁻¹ was the Si-O bending mode. The absorbance peak at 450 cm⁻¹ was the Si-O rocking mode. In the FT-IR spectrum of the SiN_y film, the absorbance peaks of Si-N stretching mode near 840 cm⁻¹ and wagging mode near 460 cm⁻¹ are predominant spectral features [5]. The main absorbance peaks in the FT-IR spectra of three SiO_xN_y films continuously shifted from that of SiN_y toward SiO_x as the content of nitrogen in the SiO_xN_y films decreased. The shape of FT-IR spectrum of SiO_xN_y1 film largely resembled that of SiO_x due to the low nitrogen concentration (5.60 at.%), while the spectra of both SiO_xN_y2 and SiO_xN_y3 films exhibited relatively broad absorbance peaks. If the SiO_xN_y films were random mixtures of SiO_2 and Si_3N_4 phases according to the RMM structure, the absorbance peaks should be completely resolved into two separate spectral peaks of Si-O (1075 cm⁻¹) and Si-N (840 cm⁻¹) stretching modes. However, such peak deconvolution could not be accomplished for the absorbance peaks of these SiO_xN_y films, indicating that the broad absorbance peaks in SiO_xN_y FT-IR

177

spectra were the combinations of overlapping peaks with different Si-O and Si-N variations in a possible Si-(O, N) random bonding environment. Therefore, the FT-IR results might reveal that co-sputtered SiO_xN_y films could be represented by the RBM structure constituted by one homogenous phase of a disordered O-Si-N network.

Nanoindentation creep on sputtered SiON films

The load – displacement (P-h) curve of the nanoindentation creep on SiON films is shown in Figure 3. For the steady-state time-dependent plastic deformation (or creep) in a uniaxial test, the relationship between the constant strain rate and applied stress can be described by an empirical equation,

$$\dot{\varepsilon} = C\sigma^n \qquad (1)$$

in which $\dot{\varepsilon}$ is the constant strain rate, σ is the applied stress, n is defined as stress exponent of the power-law creep, and C is a material related constant. The applied stress σ can be obtained as the average pressure under the indenter, or the indentation hardness,

$$\sigma = \frac{P}{A} \sim \frac{P}{h^2} \qquad (2)$$

where A is the indenter area function as $A = F(h)$, P is the instantaneous indentation load, and h is the instantaneous indentation displacement. The constant strain rate $\dot{\varepsilon}$ could be substituted by an equivalent strain rate defined as

$$\dot{\varepsilon} = \frac{1}{h}\frac{dh}{dt} \qquad (3)$$

where h is the instantaneous indentation displacement, and t is the indentation time [35-36]. The indentation displacement rate dh/dt during the creep period can be calculated by fitting the creep displacement-time curve within the constant load holding period with the following empirical law,

$$h(t) = h_0 + a(t - t_0)^b + ct \qquad (4)$$

where h_0 and t_0 are defined as the initial indentation displacement and creep initial time, respectively, and a, b and c are fitting constants [6]. By obtaining the equivalent strain rate $\dot{\varepsilon}$ and applied stress σ from the nanoindentation data by using equations (2), (3) and (4), the stress exponent n, which reveals the creep properties and mechanisms of the given materials, can be extrapolated from the slope of logarithm of $\dot{\varepsilon}$ versus logarithm of σ curve [log($\dot{\varepsilon}$)-log(σ)] plotted from equation (1).

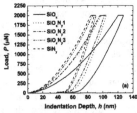

Figure 3. Load – displacement (P-h) curves of nanoindentation creep on the SiON films.

Figures 4(a)-(e) demonstrate logarithm relationships between strain rate and applied stress [log($\dot{\varepsilon}$)-log(σ)] in nanoindentation creep on respective SiON films. During the creep period, the stress decreased with creep time due to the constant applied load and the increases in both creep displacement and contact area. Therefore, the stress exponents of SiON films, i.e. the slopes of all curves [∂ log($\dot{\varepsilon}$)/∂ log(σ)], declined as stress decreased with creep time.

However, it is noted that for the SiO_x and high oxygen-content SiON films (*i.e.*, specimens SiO_xN_y1 and SiO_xN_y2), the stress exponents declined significantly from a large value ($n \sim 100$) at the start of creep to a steady-state value less than 10 at the end of the creep period. On the contrary, for the SiN_y and high nitrogen-content SiON films (*i.e.*, specimen SiO_xN_y3), the stress exponents decreased relatively slowly, from a similar starting value to a final steady-state value greater than 25.

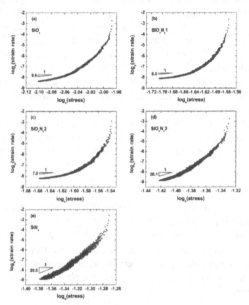

Figure 4. Logarithm relations of strain rate and applied stress [log($\dot{\varepsilon}$)-log(σ)] in nanoindentation creep for the SiON films: (a) SiO_x; (b)SiO_xN_y1; (c) SiO_xN_y2; (d) SiO_xN_y3 and (e) SiN_y.

<u>Shear transformation zone (STZ) theory</u>

The different nanoindentation creep behaviors observed in SiON films were interpreted by the shear transformation zone (STZ) theory of amorphous plasticity [7, 8]. In plastic deformation mechanism of amorphous materials, the "shear band" describes a microscopic region of localized plastic deformation under high level of stress/strain rate. The formation of a shear band is associated with the shear transformation of a small volume of material consisting of atoms (tens of atoms at most) with a certain orientation, known as a basic shear unit - "shear transformation zone". The process of shear deformation of multiple STZs determined the characteristic of plastic flow, which is represented as the stress exponent n in equation (1). A large value of n indicates that the flow is inhomogeneous and concentrates in a limited number and size of STZs. A small value of n then indicates a more homogenous and diffusional flow in plastic deformation.

Despite of the thermal annealing at 400 °C for 30 minutes, all SiON films sputtered at room temperature were still largely amorphous, and inevitably contained defects including structural voids and distorted (compressed or stretched) or defected (dangling or broken) bonds. Taking into account the characteristics of atomic bonding in the materials, the STZs clusters in sputtered SiON films upon nanoindentation creep were believed to consist of fundamental Si-O (for SiO_x), Si-N (for SiN_y) or O-Si-N (for SiO_xN_y with a RBM structure) tetrahedral units, which were bonded by the distorted and/or defected Si-O-Si and Si-N-Si bridging bonds. As demonstrated in Figures 4, at the beginning of the creep loading, the high level of stress resulted in an inhomogeneous plastic flow with a large stress exponent in all SiON films. When the applied stress exceeded the threshold stress, individual STZ clusters under the indenter containing Si-O, Si-N or O-Si-N tetrahedral units were collectively activated and started to move. This localized atomic-scale shear transformation further lead to rapid interaction, propagation and rearrangement among the nearby STZs, as well as the coalescence of adjacent free volumes. A mechanical softening in materials thus occurred as the average applied stress into the materials was reduced due to formation of relatively large shear bands of STZs with accumulated free volume. For the amorphous SiO_x film and high oxygen-content SiO_xN_y films [Figure 4(a)-4(c)], the relatively flexible Si-O-Si bridging bonds and less dense packing Si-O structure could attribute to the formation of numerous shear bands to accommodate the applied stress and strain. The stress exponent gradually declined to a steady-state value, indicating that a more homogeneous flow ($n < 10$) was eventually reached. However, for the amorphous SiN_y film and high nitrogen-content SiO_xN_y film [Figure 4(d)-4(e)], some activated STZs might be blocked by the rigid Si-N-Si bridging bonds in movement. Therefore, few shear bands were produced in these films through STZs propagation, resulting in a more localized and less homogeneous flow ($n > 25$).

CONCLUSIONS

The SiO_xN_y films deposited by co-sputtering silicon oxide and silicon nitride were composed of one homogeneous phase of random bonding O-Si-N network, *i.e.*, a structure interpreted by the RBM. Upon nanoindentation creep, the plastic flow in high oxygen-content SiON films was more homogeneous than that of high nitrogen-content SiON films. A deformation mechanism established on the chemical bonding structure and STZ amorphous plasticity theory was proposed to explain the observed variations in creep response of SiON films.

ACKNOWLEDGMENTS

This work was financially supported by the funding from the Air Force Office of Scientific Research (AFOSR) under contract number FA9550-06-1-0145 and the National Science Foundation (NSF) under contract number CMMI-0700688. The authors gratefully thank Professor Catherine Klapperich for her support with the nanoindentation instrument.

REFERENCES

[1] S. Habermehl, A. K. Glenzinski, W. M. Halliburton and J. J. Sniegowski, *Mater. Res. Soc. Symp. Proc.* **605**, 49 (2000).
[2] W. C. Oliver and G. M. Pharr, *J. Mater. Res.* **19**, 3 (2004).
[3] H. R. Philipp, *J. Non-Cryst. Solids* **8-10**, 627 (1972).
[4] G. Lucovsky, P. D. Richard, D. V. Tsu, S. Y. Lin and R. J. Markunas, *J. Vac. Sci. Technol. A* **4**, 681 (1986).
[5] D. V. Tsu, G. Lucovsky and M. J. Mantini, *Phys. Rev. B* **33**, 7069 (1986).
[6] H. Li, A. H. W. Ngan and *J. Mater. Res.* **19**, 513 (2004).
[7] A. S. Argon, *Acta. Metall.* **27**, 47 (1979).
[8] J. S. Langer, *Phys. Rev. E* **64**, 0115041 (2001).

AUTHOR INDEX

SUBJECT INDEX

Printed in the United States
By Bookmasters